U0250967

“一带一路”

服饰·语言·文化·艺术探索

李傲君　张慧琴◎主　编
刘　颖　杨武遒　张东晓◎副主编

中国纺织出版社

程应春

学术论坛现场

内容提要

本书选取北京服装学院第十三届“科学·艺术·时尚”节高层学术论坛期间，来自日本、美国、英国、德国、俄罗斯、塔吉克斯坦以及中国等十位学者的发言稿，以不同视角探索“一带一路”服饰文化发展历程。从龟兹服饰变迁、龟兹壁画和丝绸之路服饰，到中亚服饰、佛教女性头饰、古埃及服饰和丝绸之路跨文化交际，再到我国苗族传统服饰结构的智慧，以及《论语》中服饰文化礼仪等，多角度、全方位论述了服饰文化历史发展变化，为新时期“一带一路”服饰文化的探索与研究提供参考。

本书图文并茂，适合从事“一带一路”服饰文化研究与设计的学者研读，也适合广大服饰文化爱好者阅读与收藏。

图书在版编目（CIP）数据

“一带一路”服饰·语言·文化·艺术探索 / 李傲君，张慧琴主编．-- 北京：中国纺织出版社，2018.11

ISBN 978-7-5180-5123-6

Ⅰ．①一…　Ⅱ．①李…　②张…　Ⅲ．①服饰文化—研究—世界　Ⅳ．①TS941.12

中国版本图书馆 CIP 数据核字（2018）第 120178 号

策划编辑：李春奕　　责任编辑：苗　苗　　责任校对：楼旭红

责任设计：何　建　　责任印制：王艳丽

中国纺织出版社出版发行

地址：北京市朝阳区百子湾东里 A407 号楼　邮政编码：100124

销售电话：010-67004422　传真：010-87155801

http://www.c-textilep.com

E-mail: faxing@c-textilep.com

中国纺织出版社天猫旗舰店

官方微博 http://weibo.com/2119887771

北京玺诚印务有限公司印刷　各地新华书店经销

2018 年 11 月第 1 版第 1 次印刷

开本：787 × 1092　1/16　印张：9.5

字数：119 千字　定价：78.00 元

前 言

本书源于北京服装学院第十三届“科学·艺术·时尚”节高层学术论坛之成果。

“科学·艺术·时尚”节是北京服装学院全校性的科技文化学术活动，旨在展示学校办学成果、深化艺工融合的办学特色、促进学术交流、提高人才培养质量。该活动是北服校园文化的重要载体，被誉为北京服装学院最美的文化品牌。

第十三届“科学·艺术·时尚”节以“融合·丝路”为主题，集合了精彩纷呈的11个板块，具体包括：

（1）2017 IYDC国际青年设计师邀请赛。以“文化·融合”为主题，强调“一带一路”文化交流与创新，来自16个国家和地区的青年设计师们精心创作的110个系列作品，使前瞻性市场与商业转化结合，各国传统文化艺术在自我审美的理解中实现了融合和突破。从2014年开始举办至今，由北京服装学院主办，北京国际设计周、中国服装设计师协会及IFFTI（国际时装院校联盟）协办的国际青年设计师邀请赛，已经成为国际时尚院校之间相互交流，全球新锐设计师展示作品的知名平台。

（2）第七届首都大学生创意集市。由北京市委教育工委和北京团市委主办，北京服装学院承办，成为助力大学生创新、创意和创业梦想的时代舞台。本届主题为“青春创意，志蕴雄安”，征集了来自北京大学、清华大学、中央美术学院、北京服装学院、中国人民大学、北京工业大学等近50所驻京高校大学生，以及天津市、河北省部分高校大学生的上万件创新创意作品，涵盖了平面艺术、科技小发明、服装服饰、插花装饰、金工和雕塑等不同类别，精彩绝伦的创新创意盛宴成为提升学生创新意识、创新思维与创新能力的灵动载体，也使青年学生在积极参与中走近雄安，热爱雄安，助力雄安大发展。

（3）“丝绸之路”龟兹壁画艺术展。在北京服装学院与新疆龟兹研究院通力合作中拉开序幕，成为继2013年学校成功举办“垂衣裳——敦煌艺术大展”之后，又一响应国家“一带一路”倡议的壁画艺术与服饰主题融合的艺术盛会。展品从龟兹壁画数字高清图像、龟兹壁画临摹作品、龟兹石窟模拟仿真洞窟等，到北服师生的临摹龟兹壁画中的动物、植物等纹饰作品，以及用其中的元素进行的服饰设计创新，乃至师生共创的设计作品等，都是展现龟兹艺术迷人风采，传承和弘扬丝绸之路传统文化艺术，实现壁画艺术创新合作的新成果。

（4）《玉石之路》玉雕精品展。以东接中原腹地，西连中亚地区的“丝绸之路”之雏形——“玉石之路”具有代表性的十位苏作玉雕大师的作品为展品，追溯源于新石器时代的苏州玉文化，关注其“精、细、雅、巧”的特色，品味其绘画与雕塑语言，并融入创新元素的“新苏作”境界，祈愿这项在2008年就已列入国家级非物质文化遗产代表作名录的玉雕技艺，在精细雅洁中传承创新。

（5）丝绸之路艾德莱斯丝绸藏品展。艾德莱斯丝绸是古丝路多元文化汇聚交流的典型代表，其织染技艺早在2008年就被列入国家级非物质文化遗产名录，新疆艾德莱斯研发推广中心设计总监，新疆维吾尔自治区服装（服饰）行业协会副会长、新疆服装设计师协会副主席、全国十佳时装设计师程应奋老师将不同国家和地区的艾德莱斯丝绸藏品和时装作品，共计50余件参展，体现了艾德莱斯丝绸融汇中国丝绸和世界各地艾德莱斯风格的特点，其独特的民族风情与服饰文化形态，承载着不同国度和民族的情感与文化。这融合八方文化的艾德莱斯丝绸，宛如一幅幅“丝绸之路”的风景图。

（6）“一带一路”艺术交流展。此艺术交流展汇集了中国当代实力派

画家、龟兹艺术专家、敦煌艺术专家、中央美院教授、中国国家画院研究员与北京服装学院知名教授的大作，契合国家“一带一路”倡议——“融合与发展”，多角度记录了从克孜尔到敦煌、再到西安等地考察采风中的见闻，展示了西部风光、丝路遗迹与民族风情，其作品种类涵盖国画、油画与雕塑等多元面貌，呈现出一场别具特色的静态视觉艺术效果，见证了古代东西文化艺术的碰撞与交融。

（7）艺工融合科技展。艺术融合科技展是学校多学科、多领域融合协作的创新科技研发成果在五个区域的展示。“智能材料”展区展示针对空气污染防护的过滤纤维及自主研发的空气过滤装置、日常防护用口罩、新型电致变色窗、柔性变色服装和纺织品；“功能纤维和绿色纤维”展区包括拥有自主知识产权的环保型高阻燃纤维，静电纺纳米纤维膜、面料、防水透湿服装及性能展示，废旧纺织品循环利用技术和产品展示；“染整工程”展区中，数码印花无需制版，设计样稿可在计算机上任意修改。现场可转印图案，即时体验和互动；“智能生活”展区展品为功能机器人、炫彩视觉、可穿戴功能服装、创意生活家居作品等；“虚拟服装”展区将体感交互技术应用于试衣系统，使人感受到计算机交互技术发展带来的便利。系统除提供服装的试穿功能外，还借助文字、图片等形式，全面介绍不同服装所对应的文化背景，普及服装文化常识。

（8）DESIGN DAY 设计马拉松。由中国北京服装学院、澳大利亚科廷大学与韩国国民大学三家主办的 DESIGN DAY 设计马拉松，是全方位的快速设计竞赛，共分三个环节：论坛、24 小时快速设计、汇报颁奖。该设计借助创新思维来开发企业委托的设计项目，带动不同领域企业与参赛学员之间的合作，涉及的竞赛方向包含设计规划、空间设计、视觉设计、产品设计、互动设计、时尚设计等。参与马拉松的设计师须具备较强的抗压力，在24小时内竭尽所能，最终提交具体且可行性的设计方案，获胜团队将获得奖金奖品与相应的实习与工作机会。

（9）平面设计在中国北服巡展及论坛。作为中国最具设计影响力和专业荣耀的设计大奖——GDC（平面设计在中国展）打破了设计领域边界，通过无格式、无限制的互动集会，将触角延伸至设计、社会、价值等层面，重新审视当代设计行业的现状与未来，激发设计的潜在力量及深层影响力，为每一个参与者搭建自由、互动的知识分享平台。此次GDC Show以反映当下华人地区设计行业实况，促进行业交流，刷新并深化传播视觉传达的设计价值，启发设计趋向为目的，前后共有七位著名设计师分享他们的实践创作与思考，从视觉传达、时代主张、信息介质、多维呈现等不同角度来重新审视当下设计与文化、商业、生活的互动关系，涉及人文思考、美学探究、方法构建、商业创新、角色定位等设计类议题的公众分享和交流。

（10）江南三织造——李雨来藏清代宫廷服饰萃珍暨清代服饰文化研习系列活动。集中展出李雨来先生珍藏的七十余件清代宫廷服饰（均来源于江南三织造）。以明末清初、康乾盛世、清末民国三个江南丝织业发展的重要历史时期为线索，纵览宫廷服饰在结构、形制上的演变过程，再次见证了江南作为丝绸原产地，以江宁织造局（南京）、苏州织造局、杭州织造局为代表的"江南三织造"成就了我国古代丝织技术发展的高峰。同时，北京服装学院研究生院以中国传统服饰文化抢救传承与设计创新博士项目为依托，邀请学者专家就以清代服饰文化为代表的中国古代服饰文化的最新研究成果作公开讲演，组织博士、硕士研究生对古代服饰文化传承与创新研究的方法和思路展开讨论，清代服饰文化研习系列活动也在理论与实践的融合中吸引了无数学子的参与。

（11）“一带一路”服饰文化、艺术、产业高端学术论坛。论坛共分为两个版块，产业版块与服饰文化、艺术版块。产业版块举办的活动汇聚政府、行业、企业和学术智慧推动服装纺织业的发展，在营造多元文化碰撞的学术交流氛围中，促进文化认同。从工业和信息化部消费品工业司纺织处曹庭瑞处长“把握‘一带一路’机遇，促进纺织工业升级”的报告，到中国贸促会研究院国际贸易部主任、研究员，兼任商务部特聘“全国内贸行业专家”赵萍围绕“一带一路”带来的贸易投资新机遇的报告；从国家信息中心大数据发展部大数据分析处负责人，国家信息中心“一带一路”大数据中心杨道玲主任助理针对国家“一带一路”大数据决策支持体系建设与思考的报告，到中国社会科学院亚太与全球战略研究院（原亚洲太平洋研究所）区域合作研究室主任，中国社科院研究生院博士生导师王玉主教授所作的“一带一路”倡议：背景与前景；从中国纺织工业联合会国际贸易办公室贸易政策处处长，中国国际贸易促进委员会纺织行业分会贸易投资促进部刘耀中主任对于“一带一路”的纺织布局与展望，再到融天工（北京）文化发展有限公司董事长，北京捷盟管理咨询有限公司首席顾问罗先初，就服装产业转型升级与传统工艺传承创新展开的论述与剖析，使师生对于“一带一路”有了全方位、立体的认识与理解，为学科建设与专业发展提出了新思路，开辟了新路径。

同时，在“一带一路”服饰文化、艺术、产业高端学术论坛的服饰文化、艺术版块的活动中，邀请了北京服装学院贺阳教授等国内学者以及来自日本、美国、英国、德国、俄罗斯、塔吉克斯坦的学者，共计十三名，从不同视角探索龟兹服饰变迁、龟兹壁画、丝绸之路服饰、中亚服饰、佛教女性头饰、古埃及服饰、丝绸之路跨文化交际，我国传统苗族服饰结构的智慧，《论语》中服饰礼仪的哲学层面思辨，“一带一路”

沿线各国语言消费，以及英语语言中基于定冠词的使用，剖析语言文化思维差异等。

上述对于“科学·艺术·时尚”节的介绍主要选自于北京服装学院宣传部网站信息，而本书的形成则是源于“一带一路”服饰文化、艺术、产业高端学术论坛的服饰文化、艺术活动中十一位专家讲座的初稿。虽然有两位专家的讲稿因时间紧急未能如期收录，书中讲稿具体排序和内容与当时的讲座略有出入，但是总体而言，还是较为完整地以书面形式再现了当时专家的发言，不妥之处还望诸位专家与读者理解和海涵。

在加强“丝绸之路”经济带建设，扩大中西部开放，打造中西部经济升级版主引擎的今天，高校的发展同样需要对接“一带一路”倡议，落实京津冀一体化协同发展，围绕建设北京“世界城市”“时装之都”“设计之都”多位一体的宏伟蓝图，立足传统文化底蕴，思考现代设计理念，深化政、产、学、研、用彼此之间的相互融合，坚持以和平合作、开放包容、互学互鉴、互利共赢为核心的丝路精神，携手推动“一带一路”建设行稳致远，将“一带一路”建成和平、繁荣、开放、创新、文明之路，迈向更加美好的明天。

张慧琴

2018年1月10日

樱花东街甲2号

目 录

Guzel Maitdinova

俄罗斯－塔吉克（斯拉夫）大学教授

丝绸之路文明化结构中的塔吉克斯坦服饰

The Costume of Tajik People in the Civilizational Structure of the Silk Road / 15

Ines Konczak-Nagel

慕尼黑大学佛教研究博士，莱比锡撒克逊科学院研究员

从新角度探讨5-7世纪龟兹壁画中吐火罗王国的佛教女性头饰

The Headdress of Mythological Female Figures in the Buddhist Mural Paintings of the Tocharian Kingdom of Kucha (5th–7th Century) — Another Point of View / 27

Josef Mueller

英国伦敦摄政大学语言与文化学院院长

"一带一路"走廊沿线的跨文化交流

The Intercultural Relations and Communication Along the One Belt One Road Corridor / 41

Li Qimei

QML文化咨询公司管理者

穿行于丝绸之路

The Silk Road, the History / 47

Stuart Walker

英国可持续设计的先锋学者，英国兰卡斯特大学想象力研究中心创始人和管理者，英国肯辛顿大学客座教授，加拿大卡尔加里大学名誉教授

为生活设计：在抽象世界中创造意义

Design for Life: Creating Meaning in a Distracted World / 53

Sergey A. Yatsenko

俄罗斯国立人文大学历史和文化理论学院教授

5-8 世纪丝绸之路服饰研究的些许问题

Some Problems of the Silk Road Costume Studies for the 5th-8th Century / 73

Inoue Masaru

中亚丝绸之路壁画的著名研究专家

龟兹石窟壁画的供养人和 6 世纪中亚及中国的服饰

Donor Figure of Kucha Murals and Costume of Central Asian and Chinese in 6th Century / 91

程应奋

新疆艾德莱斯研发推广中心设计总监，中国十佳时装设计师，新疆十佳服装设计师

艾德莱斯——非遗活在当下

Atlas — the Living Intangible Culture / 103

贺阳

北京服装学院民族服饰博物馆馆长，博士生导师

传统苗族服饰结构中的智慧

Wisdom in Traditional Costume Structure of the Miao Ethnic Group / 115

张慧琴

北京服装学院语言文化学院院长，硕士生导师

《论语》中以“礼”服人的哲学思辨

Confucius' Thoughts Reflected in the Function of Clothes in *the Analects of Confucius* / 123

李艳

首都师范大学北京语言产业研究中心执行主任，博士，教授

语言消费：基本理论问题与亟待搭建的研究框架

Language Consumption: Basic Theoretical Issues and Research Framework / 131

后记 / 149

作者简介 / 151

个人简介

Guzel Maitdinova

俄罗斯－塔吉克（斯拉夫）大学教授

历史学博士，获艺术评论学位，现任俄罗斯－塔吉克（斯拉夫）大学教授，塔吉克斯坦建筑学院通信会员。拥有苏联（Union of Soviet Socialist Republics, USSR）国家发明创造委员会15项原创专利，2项塔吉克斯坦共和国创造专利权。发表超过180项研究作品。发表过的专著包括：《吐火罗斯坦服饰：历史和联系》（1992）、《塔吉克斯坦古代和中世纪时期肉食消耗文化和体系》（与A. E. 内玛提合著，1993）、《中亚中世纪早期面料》（1996）、《塔吉克斯坦古代和中世纪早期服饰》（2003）、《塔吉克服饰史》（2卷，2004）、《俄罗斯经济利益在塔吉克斯坦地区面临的机遇与挑战》（与G. V. 科什拉科夫，M. 图拉娃合著，2009）、《丝绸之路中亚地区文明对话：历史融合和21世纪新坐标》（2015）等。另外，还曾就国际关系和地理政治问题发表多篇学术论文。

Guzel Maitdinova is the Doctor of History, the Candidate of Art criticism, Professor of the Russian- Tajik (Slavic) University, and the Corresponding Member of the Tajik Academy of Architecture and Construction. She owns 15 author's certificates of State Invention Committee of USSR, and 2 invention patents of Republic of Tajikistan. She published over 180 research works. The published monographs include: "The costume of Tokharistan: history and connections" (1992), "The system and the culture of meal-consumption of Tajiks—Ancient and Medieval periods" (co-authored with A.E. Negmati, 1993), "Early Medieval textiles of Central Asia" (1996), "Genesis of the Tajik costume: Antiquity and Early Middle ages" (2003), "History of Tajik costume" in 2 volumes (2004), "The Economic interests of Russia in Tajikistan: the challenges and possibilities" (co-authored with Koshlakov G.V., Turaeva M., 2009), "The dialogue of the civilizations in Central Asian area of the Great Silk Road: the experience of the historical integration and the landmarks of the 21st century" (2015). There are academic research papers on the problems of International relations and Geopolitics being published for the past decades.

丝绸之路文明化结构中的塔吉克斯坦服饰

The Costume of Tajik People in the Civilizational Structure of the Silk Road

The Great Silk Road gave a serious impulse to the costume art development of the peoples from Central Asia, the southern section of which began to function actively from 112 BC, when the Chinese Emperor Udi established diplomatic relations with Bactria. The suit of Tajiks was formed on the heritage of ancient civilizations, enriched by the achievements of cultures of different peoples as a result of cultural dialogue on the routes of the Great Silk Road.

This report focuses on reflecting intercultural contacts in the costume of the population of the southern region of Central Asia during the classical development of the Great Silk Road (1st–10th centuries). The history of the costume shows that ethnic and cultural interactions clearly manifest in it. Throughout the history of the Tajik costume, it reflected the autochthonous layer and the influence of intercultural communications. In the Central Asian region, since ancient times, there have been intense contacts not only between the agricultural and nomadic world of the region, but also distant cultures that acted as factors in the development of the costume. Poly-ethnicity, intensive trade, cultural and information exchange repeatedly in history became a resource not only for socio-economic development and a guarantor of interreligious tolerance, but also contributed to the internationalization of the costume. The formation of the traditional costume was influenced by the intercivilizational dialogue that was observed at all stages of the history of the Central Asian peoples, which contributed to the active translation of the cultural achievements of the surrounding world. Moreover, it was not a direct assimilation of cultural trends from outside, but the selection of those cultural phenomena was close in spirit. People's clothing has always preserved elements common to all peoples living in Central Asia, bearing the features of ethnicity that have developed in the process of constant mutual enrichment. The stability and sustainability of the civilizations of Central Asia was facilitated by the fact that along with the steppe culture and values of the agricultural world, the cultures of the Confucian, Indo-Buddhist, Christian, and many regions of the Islamic world were assimilated, by nature always being international. The latter circumstance was an important element in the balance of cultures and contributed to the birth of a completely new form of costume, which accumulated in itself the urban culture of peoples. Internal values and traditions of folk clothes originated from the depths of centuries. In Central Asia, local relicts, art and craft practices of the nomadic and agricultural world, traditions of the previous period and cultures of peoples along the Great Silk Road were integrated into a single new formation. Here they received their final imaginative completion, the system in the traditional costume of the peoples inhabiting the region. Already the ancient societies of Central Asia and the first civilizations formed a multi-polar cultural world with its own stable traditions, sufficiently clear in the context of ethnic, political, commercial, cultural interaction within the coexisting centers of civilization. All these phenomena contributed to the formation of ethnic features of the national costume. Therefore, there is so much in common in the traditional costume of Tajiks and other peoples of Central Asia. Continuity and traditionalism is the basic genetic factor of the Tajik costume (Fig.1).

Fig. 1 Female Costume 1st–2nd c. AD. Bactria

From the second half of the 1st millennium BC, inter-regional trade relations between East and West are being strengthened.

Active civilizational ancestors of the Tajik people—the Bactrians and Sogdians played a role on the Great Silk Road. The costume complex is of great importance in the structure of the culture and interchange of the Great Silk Road, since the materials, ornaments, cosmetics, basic forms of clothing used were not only the subjects of trade, but also were the translators of the cultural achievements of peoples from East to West. The structure of the costume reflected that active dialogue of civilizations, which contributed to the formation of tolerance and openness to the external influence of cultures of peoples in the Eurasian space (Fig.2).

Dialogue of civilizations on the routes of the Great Silk Road was reflected in the costume of the population of Sogd and Bactria-Tokharistan. The interaction of cultures was reflected in the components of the costume, on the used fabrics, ornaments, clothing, etc. The inclusion of huge integrated spaces in the ancient and medieval empires of Eurasia, the spread of world religions along the Silk Road routes, contributed to the formation of the concept of "fashion" in a suit. On the roads of the Silk Road, similar fabrics, sewing techniques, changing clothing patterns are spreading, and the dominant religions bring their own influences, reflected primarily in the decor and proportions of clothing (Fig. 3).

Fig. 2 The cotton pants. Kurgan. Tokharistan

Fig. 3 The ladies and the children costumes. Kurgan. Tokharistan. End of 4th–5th c. AD

The traditional costume of the Tajiks, as it appeared in the early twentieth century, contains a lot of ancient strata. One of the most ancient features that have survived to this day is the form and constructive solution of the classical ethnographic costume complex—outer clothing (such as dressing-gown, caftan), shirts of tunic cuts, loincloth with an inset for step and pointed headdress or a tightly fitting cap with rounded top, in many Eurasian peoples. In the 2nd millennium BC in Central Asia, it was already a fairly traditional costume ensemble. For the first time this costume complex in full complement appeared in the ensemble of clothes from Xinjiang more than four thousand years ago, although tunic-like tunics have been already found in the materials of the more ancient Xinjiang. Relics of archaic garments were preserved among archaeological finds from the Kurgan of the 1st–5th centuries (Fig.4).

A distinctive feature of Cherchen's outer clothing is the presence of a structural seam along the middle of the mill, which is probably connected with the width of the machine on which fabrics of a certain size were produced. It was in the P millennium BC in archaeological materials that the ancient Iranian attire. kandiz (cloak with false sleeves) was fixed in the ensemble of clothes from Xinjiang. The fashion for these clothes was so strong that from the middle

Fig. 4 The early medieval male costumes of Xinjiang

of the Century it spread along the early Eurasian routes in Attica, where it tranformed into a light cloak. Later, a similar type of clothing is traced in Scythians, Byzantium, and medieval Armenia. In the Central Asian costume ensemble, the cloak can be traced throughout the history of the costume: it was reflected in the early medieval painting of Tokharistan, Sogd. The cloak given in a medieval miniature art, in a traditional Tajik costume, was transformed into a veil. It is generally accepted that folk costume is a solid ethnic sign, especially for antiquity and the Middle Ages. While keeping ethnic features in a suit, the population along the highway of the Great Silk Road synchronously spread fashion for certain forms of clothing. With the preservation of ethnic characteristics, adapted to climatic conditions and spiritual needs of the population along the great highway, innovations were primarily reflected in the overall decorative solution of the costume, silhouette clothes, used artistic fabrics, and in ornaments. Dialogue of the arts on the routes of the Great Silk Road is reflected in the jewelry of Bactria-Tokharistan and Sogd, in which the syncretic traditions of peoples along the highway are clearly expressed (Fig.5).

Silk fabrics were used in the costume of the peoples of Central Asia back in the 2nd millennium BC, but it was the intensification of interaction along the

Fig. 5 Frontal image of an armored warriors in a crown with the raised right hand. 5th c. AD

Silk Road that promoted the wide spread of silk in the region. From China, raw silk and dyes came, but over time, in the south of the Central Asian region, their own silk-weaving traditions were being formed. And the traditions of silk-weaving developed synchronously throughout the region. Then in Central Asia, a number of local schools of artistic silk-weaving appeared. Silk-weaving schools were established in Tokharistan, Sogd, Ferghana, Khorezm, in the Lower Syrdarya, Turfan, Kucha, Khotan, etc. which were closely connected(Fig.6).

Silk weaving was promoted by the increase in the role of silk in interstate exchange and the increased popularity of art fabrics among the population. These art schools were closely associated with the traditions of China, Iran, Byzantium, India. This was evidenced by the inclusion of the motifs of Chinese, Sasanian, Indian, Byzantine textile art in the ornamental traditions of Toharistan and Sogd, not to mention the proximity of technological methods of weaving (Fig.7).

In the ornamentation of popular silks, ancient Iranian motifs of winged horses, mountain rams, geometric figures, plant motifs and images of the "tree of life" were used. The highest achievement of the weaving art—the production of polychrome silks, was closely related to the art of China, Iran and the peoples of Central Asia. Findings from Astana and Mount Mug (Sogd), collections

Fig. 6 The male costumes. Dilberjin. Tokharistan. 5th c. AD

Tokharistan n. 吐罗火斯坦

Ferghana n. 费尔干纳（乌兹别克斯坦地名）

Khorezm n. 花拉子模（乌兹别克斯坦地名）

Syrdarya n. 锡尔达拉（乌兹别克斯坦地名）

Turfan n. 吐鲁番

Khotan n. 和田

of Sogdian fabrics in museums testify to the commonality of the traditions of artistic, which is testified by the mass finds of cotton clothes in the burial of the ordinary population of the Old Termez, mentioned above in Kurgan (Fig.8).

Fig. 7 The female and male costumes. Balalyktepa. Tokharistan. 6th c. AD

Fig. 8 The male costumes. Penjikend. Sogd. 7th c. AD

The Great Silk Road was a system of mutual exchange in the field of fashion art. The imported exotic fabrics, clothes, ornaments gave impetus to their borrowing and creative approach to their production in a new place, they served as a stimulus for new innovations and promoted the understanding of other cultures. In the early middle ages on the slopes of the Great Silk Road, the fashion for trapezoid or hourglass shapes was widely spread, the silhouette of clothing used for silk and cotton textiles of dense texture with a large stain-forming silhouette. It was the development of communications that facilitated a broad exchange of artistic culture on the East-West routes, which in turn led to the spread of fashion. It was in the middle ages with the flourishing of the Great Silk Road in Central Asia that the formation and development of fashion in the current understanding took place. Fashion is vividly manifested in the elite layers of society, often eroding ethnic specifics in a suit (Fig. 9、Fig. 10).

The oldest forms of clothing, the principles of their decoration, which were influenced by innovations on the Great Silk Road, so far have been preserved in the traditional costume of the Tajiks.

Fig. 9　The textile pattern: the peacocks. Afrosiab. Sogd. 7th c. AD

Fig. 10 Male and female costumes. Afrosiab. Sogd. 7th c. AD

In the traditional costume complex of the Tajiks, to this day, tunic-like shirts, loincloth with a step-inset, the manner of wearing several dresses, the use of large stain-forming ornamental motifs (chikan-embroidery) with dressing of women's clothing, etc. remaining Syncretic local traditions of the historical costume of the ancestors of Tajiks - Bactrians and Sogdians, become an integral part of the ethno-cultural structure of civilizations of Central Asia (Fig.11~Fig.14).

(1) a young woman of Darvaz (2) a Darvazi maiden (3) a young woman of Kulyab

Fig. 11 The traditional costumes

(1) Karategin area (2) Oal’ ai Khumb

Fig. 12 The traditional costumes of young women

(1) Karategin (2) Kulyab (3) Darvaz

Fig. 13 The traditional costumes

Fig. 14 The traditional costumes of the bride and the groom. Ishkashim. Pamir.

个人简介

Ines Konczak-Nagel

慕尼黑大学佛教研究博士，莱比锡撒克逊科学院研究员

莱比锡撒克逊科学院研究中心研究员，2016年以来研究丝绸之路北部的库车佛教壁画，她在柏林自由大学获得印度艺术史和语言学硕士学位，2014年她在慕尼黑路德维希马克西米连大学获得佛教研究博士学位。她的研究主要集中在丝绸之路北部壁画图案的起源、传播和变化等方面。在2012年，她获得了京都龙科大学的奖学金，随后她获得了马克斯－普朗克学院，弗洛朗斯昆斯特历史学院“博物馆中的艺术史连接”项目的两年博士后研究金。她还曾在亚洲艺术博物馆、柏林国家博物馆和莱比锡大学任职。

Ines Konczak-Nagel is a research associate at the research centre of the Saxon Academy of Sciences in Leipzig "Buddhist Murals of Kucha on the Northern Silk Road" since 2016. She received her MA in Indian art history and philology from the Free University of Berlin and her Ph. D. in Buddhist Studies from the Ludwig Maximilian University of Munich (2014). Her research focuses on aspects of origin, transmission, and alteration of pictorial motifs in the mural paintings of the Northern Silk Road. In 2012, she was awarded a research fellowship at the Ryukoku University in Kyoto. Afterwards she won a two-year postdoctoral research fellowship in the program "Connecting art histories in the museums" from the Kunsthistorisches Institut in Florenz, Max Planck Institute. She has also held positions at the Asian Art Museum, National Museums in Berlin and the Leipzig University.

从新角度探讨 5-7 世纪龟兹壁画中吐火罗王国的佛教女性头饰

The Headdress of Mythological Female Figures in the Buddhist Mural Paintings of the Tocharian Kingdom of Kucha (5th–7th Century) — Another Point of View

The females depicted in the Buddhist mural paintings of the ancient Tocharian kingdom of Kucha (also called Kizil, koutcha. 5th–7th) can be divided in two distinctive types: firstly, female donors representing contemporary local supporters of the Buddhist order, and secondly, mythological figures representing persons from Buddhist legends, such as goddesses, queens, princesses, or courtesans. Both types generally differ not only in their garments but also in their headdresses. While the female donors are represented wearing indigenous fashion of the Kucha people[1], the appearance of female mythological figures is completely different and without any analogy to the local fashion. The headdresses of mythological females in particular are very diverse and generally highly elaborate.

Since no textual sources on headdresses of mythological figures were transmitted, it is only art history which can help to evaluate which kind of headdress the artists attribute to mythological figures. An art-historical examination of female headdresses in the Buddhist art of Kizil has already been undertaken by Inoue (2015). In his study, the author describes different types of headdresses of female mythological figures and traces their origin back to the art of Gandhara.

---

[1] For a study of the garments of donors in the mural paintings of Kizil, see for example Ebert (2006).

In the current paper, however, I would like to introduce another point of view concerning what the original shape of some headdresses might have looked like, and I would like to discuss from which region these types of headdress could have been transmitted.

The headdresses of the mythological female figures are by far more glamorous than those of the female donors. There is one main type of female headdress that appears in paintings of the First Indo-Iranian style (Fig. 1). This type is composed of a bun which is apparently worn on the back of the head and fixed by enclosing strings of pearls which are braided with the plait of hair in the back. The strings of pearls are held together by a kind of clasp that is set next to a large central ornament placed directly over the forehead. According to Inoue (2015), this main headdress type originated in ancient Gandhara (Fig. 2). There are some similarities to the art of Gandhara, such as the back of the head being covered by a fabric, some long curls falling onto the shoulders, and a central floral ornament above the forehead decorating a laurel wreath in the Gandharan example.

However, if we compare representations of a subject that is known to the art of Kucha as well as to the art of Gandhara, differences between both art

Fig. 1

schools become obvious. An example is the popular representation of Queen Māyā giving birth to prince Siddhārtha who later became Buddha Śākyamuni (Fig. 3). One obvious difference between both art schools is the rendering of Queen Māyā's garment. While in the Kuchean example, she is shown in a sheer transparent garment which only covers the lower part of her body, in the

Fig. 2

Fig. 3

Gandharan relief it is a comparatively heavy garment which covers her whole body. Moreover, the female headdresses are different in the two schools of art. In the Gandharan example, the ladies wear a laurel wreath with a flower ornament which is either placed in the centre or slightly laterally. In the Kuchean example, however, apart from the fact that the ladies do not wear a laurel wreath at all, the flower is not the central ornament but placed to one side of the central ornament.

This type of headdress generally appears in the paintings of Kizil Cave 76 (Pfauenhöhle, Peacock's Cave; Fig. 4). It is characterised by four main features: (1) a multi-row string of pearls that encloses the entire headdress and is braided with the plait of hair in the back, accompanied by centrally placed jewels above the forehead which are crowned by a semi-circular ornament; (2) a bun covered with a patterned textile; (3) a decorative floral ornament attached to one side of the central ornament; (4) two pendants hanging from the headdress—one attached to the hair bun and the other to the decorative flower ornament. Sometimes the ladies additionally wear a large scarf over their head whose ends fall around their elbows.

The female headdresses in the paintings of Kizil Cave 76 differ in their composition from other female headdresses of the First Style paintings of Kizil. Compared to paintings of Kizil Cave 207 (Fig. 5), for example, the headdresses might be interpreted as misunderstood representations. One might assume that the painters misunderstood the floral element that in the painting from Kizil Cave 207, apparently functions as a clasp for the framing strings of pearls, by rendering it into a floral ornament in Kizil Cave 76, seemingly made of textile, instead of the strings of pearls, a pendant emerges. The fact that in the paintings from Kizil Cave 76, the hair bun appears not to be at the back of the head as in the paintings from Kizil Cave 207 but slightly laterally offset could be explained with painters' missing sense for perspective. The bun is shown in a perspectively correct manner, located right behind the central ornament, the paintings from Kizil Cave 207. In the paintings from Kizil Cave 76, however, it is shown next to the central ornament, giving the impression that the bun is worn not at the back of the head but laterally.

Fig. 4

Fig. 5

However, what if it's not actually a case of faulty perspectival rendering by the painters? What if the viewers were wrong in the assumption that the painters were not able to render perspective?

Sometimes it is advantageous to take a different point of view in seeing something, since the image we see is created by our brain based on our knowledge and our past experiences. As such, first of all, we tend to see things how we assume them to be. This is even more true for two-dimensional objects such as paintings. Until now, we have assumed that the bun is placed symmetrically on the back of the head behind the central ornament. However, if we assume that the painters were able to render perspective, then the headdress would be an appealing asymmetrically worn arrangement.

Asymmetrically worn female headdresses are known from Kuchean paintings, thus for instance from Kizil Cave 118 (Hippokampenhöhle, Cave of the Hippocampi; Fig. 6). To the right of the main figure of King Māndhātar[1], as seen by the viewer, a lady is depicted offering the king her breast to suckle him.

---

[1] For the identification of this mural painting from Kizil cave 118, see Hiyama (2010); Hiyama (2012).

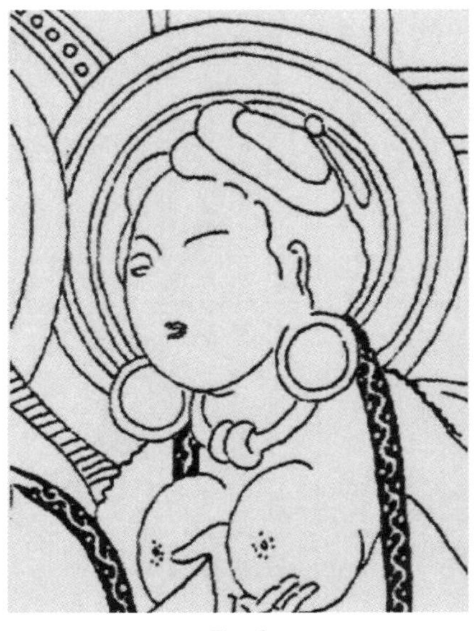
Fig. 6

Fig. 7

According to Hiyama (2012), this is certainly an allusion to the name of King Māndhātar which, according to the legend, derives from Sanskrit words that were spoken by all ladies who saw the king when he was a small child. They said "*māṃ dhātu*" which means "Let him suck me!"[1]

The lady who offers King Māndhātar her breast wears something on the left side of her head that is reminiscent of a small cap. A somehow similar, asymmetrically worn headdress can be found in a painting from Kizil Cave 84 (Schatzhöhle B, Treasure Cave B), although in this case the composition is much more complex and includes dangling strings of pearls (Fig. 7). A closer look reveals the composition of this headdress. It is a little hair knot which is tied together with a small string and surrounded by a broad ring possibly made of textile.

A female headdress that is reminiscent of this kind of composition is known from the Swat region (Fig. 8). In the relief, it is clearly visible that it is a laterally worn plait of hair framed by an ornamental ring. It is important to

[1] For the corresponding source references, see Hiyama (2012: note 19).

Fig. 8

note that although the Swat region geographically belongs to Gandhara, the art of this region is much closer to Central India than to the typical Hellenistic art of Gandhara. Indeed, the tradition of asymmetrical worn headdresses is well known in the art of the Indian subcontinent and was already very popular during the Śuṅga period from the 2nd to the 1st century B.C.E. [1]

Headdresses worn in two parts were also in fashion in India during the Śuṅga period. In a couple of terracotta figures from Kaushambi, a lady is shown in frontal view wearing a bun surrounded by small flowers on the left side of her head, and three large flowers in her hair on the right side.[2] The whole composition is framed by two parallel running strings of pearls which are further decorated with one flower on each side.

---

❶ For examples of East Indian terracotta figures from the Śuṅga period showing females with asymmetrically worn headdresses, see Haque (2001: 108–111, 182).

❷ Some of these terracotta figures are kept in the National Museum New Delhi, see for example, http://www.museumsofindia.gov.in/repository/record/nat_del-90-103-21358; some are kept in the Alahabad Museum, see for example, http://www.museumsofindia.gov.in/repository/record/alh_ald-AM-TC-K2560-336 (2017, October 31).

There are other examples of two-part headdresses from the Śuṅga period where the two parts clearly differ in size. An East Indian terracotta from Bengal [Fig. 9(1)] shows a female in frontal view wearing a small bun covered by a patterned textile on the left side of her head and possibly pins wrapped by an ornamental band on the right side. Here too, the entire composition is framed by parallel running strings of pearls.

In some points, the Bengal terracotta is similar to Queen Māyā from Kizil Cave 76 [Fig. 9 (2)]. Not only do both figures stand in a similar posture, they also wear similar dresses with a multi-row girdle consisting of small plates. Moreover, the headdresses of both ladies share some features. They consist, in one part, of a bun covered by a patterned textile while the whole composition is framed by a multi-row string of pearls.

The tradition to arrange women's hair in two parts of different size is equally known from the Swat region. One example from Butkara (Fig. 10) shows a lady in frontal view wearing a small hair knot on the right side of her head

(1)

(2)

Fig. 9

Fig. 10

and a bigger knot, covered by patterned textile, on the left side. Similar to the painting from Kizil Cave 76, the lady wears a scarf over her head, with the ends falling around her elbows.

As we have seen, the tradition of asymmetrically worn two-part headdresses was well-known in ancient India and could have influenced the art of Kucha via the Swat region. However, there still is one feature that cannot be explained by the examples of Indian art that we have seen so far. This feature is the pendants dangling from the headdress.

Representations of women's headdresses which in terms of pendants, might be comparable to the headdresses of the ladies from Kizil cave 76 can be found in the so-called Begram ivories[1]. The style of the ivory objects discovered in Begram is very close to the art of Mathura in Central India and Amaravati in

[1] These ivories once covered precious furniture, such as sofas and chairs. They were found in two rooms which are supposed to were storage rooms of a wealthy merchant who could have collected such furniture. See Tissot (2006: 134).

South India. For this reason Tissot (2006) assumes that the ivory objects were imported from these regions. As a matter of fact, such small portable objects are very suitable for spreading pictorial motifs.

In some respects, the headdresses of the ladies of the Begram ivories (Fig. 11) are quite similar to those of the ladies from Kizil Cave 76. Both headdresses are asymmetrically arranged with a laterally worn bun covered by patterned textile, and both are provided with two pendants. In the Begram ivories, one of the two pendants hangs down in a loop from the hair bun and might be of textile or hair, while the second one hangs down from the other side of the head and might represent a small branch of an Aśoka tree. In this respect, the headdresses are similar to those worn by the ladies from Kizil Cave 76. On one side of the head, there is a bun with an attached pendant, and on the other, a botanical pendant or flower with an attached pendant. What is lacking in the Begram ivories is the central ornament and the framing string of pearls. However, there may have existed similar Indian depictions with a central ornament which served as a model but is no longer preserved today.

The idea of an asymmetrically composed headdress with lateral parts also exists in the paintings of the Second Indo-Iranian style. In an example from Kizil Cave 80 (Fig. 12), the lady wears an angular cap on the right side of her head and a small bun on the left side. A bigger bun is worn on top of her head in front of which a central ornament with a flower on top is placed. The same type can be found depicted in a number of other caves in Kizil only the the angular cap is sometimes replaced by a roundish bun, as for instance in Cave 199 (Fig. 13).

On the whole, there are differences in the representation of female headdresses in the First Style paintings from Kizil. For example, in Cave 207, the dominating bun of the headdress is unmistakeably depicted on the back of the head, whereas in Cave 76, it is painted on one side of the head. The reason for this distinct rendering is not necessarily the painters' inability to render perspective. It is more probable that the painters used different models for their paintings. The female headdress in paintings from Kizil Cave 207 seems to

Fig. 11

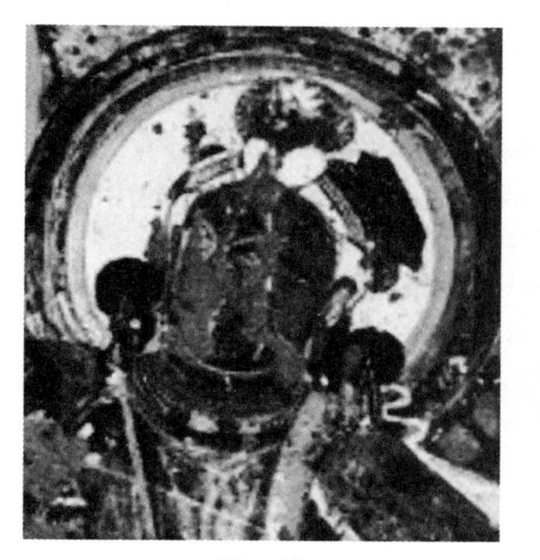

Fig. 12

Fig. 13

derive from representations of females in the Graeco-Buddhist art of Gandhara. This is especially indicated by the small tuft of hair on the back of the head.

The female headdress in paintings from Kizil Cave 76, however, seems to derive from representations of females in ancient Indian art. This is especially indicated by the asymmetrical composition which is known to Indian art since at least the 3rd century B.C.E. and was still in vogue in India in the 5th century C.E. as can be seen in paintings from the Buddhist cave temples in Ajanta in Central India[1].

---

[1] One example of a depiction of a female wearing an asymmetrical headdress is the so-called "black Apsaras" from Ajanta cave XVII. https://en.wikipedia.org/wiki/Ajanta_Caves#/media/File:Ajanta_Cave_17_veranda_mural_beauty.jpg (2017, October 31).

Bibliography

[1] Albanese, M. Das antike Indien: von den Ursprüngen bis zum 13. Jahrhundert[M]. Cologne: Müller, 2001.

[2] Ebert, J. The Dress of Queen Svayaṃprabhā from Kuča: Sasanian and Other Influences in the Robes of Royal Donors Depicted in Wall Paintings of the Tarim Basin. In: Schorta, R. (ed.): Central Asian Textiles and their Context in the Early Middle Ages[J]. Riggisberg: Abegg-Stiftung, Riggisberger Berichte 9, 2006, 101–116.

[3] Faccenna, D. Sculptures from the Sacred Area of Butkara I (Swat, W. Pakistan) [M]. Roma: Istituto Poligrafico dello Stato, Libreria dello Stato, Reports and memoirs, Istituto Italiano per il Medio ed Estremo Oriente, Centro Studi e Scavi Archeologici in Asia 2, 1964.

[4] Grünwedel, A. Altbuddhistische Kultstätten in Chinesisch-Turkistan: Bericht über Archäologische Arbeiten von 1906 bis 1907 bei Kuča, Qarašahr und in der Oase Turfan [M]. Berlin: Reimer, Königlich Preussische Turfan-Expeditionen,1912.

[5] Grünwedel, A. Alt-Kutscha: Archäologische und religionsgeschichtliche Forschungen an Tempera-Gemälden aus Buddhistischen Höhlen der ersten acht Jahrhunderte nach Christi Geburt [M]. Berlin: Elsner, Veröffentlichungen der Preußischen Turfan-Expeditionen, 1920.

[6] Haque, E. Chandraketugarh: A Treasure-House of Bengal Terracottas [M]. Dhaka: The International Centre for Study of Bengal Art, Studies in Bengal Art Series 4, 2001.

[7] Hiyama, S. Kizil Sekkutsu dai 118 kutsu (Kaibakutsu) no Hekiga Shudai: Mandataru ou Setsuwa wo Tegakarini (The Wall Painting of Kizil Cave 118: The Story of King Māndhātar as a New Identification) [J]. Journal of Art History, 2010: 168, 358–372 (in Japanese), 6–7 (English Summary).

[8] Hiyama, S. A New Identification of the Murals in Kizil Cave 118: The Story of King Māndhātar [J]. Journal of Inner Asian Art and Archaeology 5, 2012: 145–170.

[9] Inoue, M. Women's Headdress Depicted in Mural Paintings of Kizil: New Perspectives on the Silk Road. Silk Road: Meditations. 2015 International conference on the Kizil Cave paintings, collection of research papers [C]. Shijiazhuang: Hebei Fine Arts Publishing House, 2015.

[10] Kurita, I. Gandharan Art. 2 Vols, Ancient Buddhist Art Series [M]. Tokyo: Nigensha Publishing, 2003.

[11] Le Coq, A. von. Die buddhistische Spätantike in Mittelasien IV: Atlas zu den Wandmalereien. Graz: Akademische Druck- und Verlagsanstalt, Ergebnisse

der Kgl. Preussischen Turfan Expeditionen[M]. Berlin: Fines Mundi GmbH Saarbrucken, 1974.

[12] Mehendale, S. The Begram Carvings: Itinerancy and the Problem of "Indian" Art. In: Aruz, J. and Fino, E. V. (eds.) Afghanistan: Forging Civilizations Along the Silk Road[M]. New Haven: Yale University Press, 2012.

[13] Tissot, F. The Art of Gandhâra: Buddhist Monk's Art, on the North-West Frontier of Pakistan[M]. Paris: Libraired' Amérique et d' Orient, 1986.

[14] Tissot, F. Catalogue of the National Museum of Afghanistan: 1931 – 1985[M]. Paris: UNESCO Publishing, Art, Museums and Monuments Series, 2006.

[15] 中国新疆壁画艺术编辑委员会，中国新疆壁画艺术克孜尔石窟 [M]. 乌鲁木齐：新疆美术摄影出版社，2009.

个人简介

Josef Mueller

英国伦敦摄政大学语言与文化学院院长

现任英国伦敦摄政大学语言与文化学院院长。这个学院包括一个英语语言中心，开设九种外语课和文化间交流。出生在奥地利，在英国从教25年。教授的课程包括跨文化交际、跨文化管理、文化与身份、国际学生合作、德语等。他曾担任过摄政大学工商管理系副主任，在诺丁汉大学教授德语，他的研究包括文化身份与跨文化适应。

Josef Mueller is currently Director of the Institute of Languages and Culture at Regent's University London, UK. The Institute comprises an English Language Centre and offers tuition in nine foreign languages, as well as in intercultural communication. Originally from Austria, he has been living and working in the UK for 25 years. His teaching comprises courses on intercultural communication, cross-cultural management, culture and identity, working with international students, as well as German language. His previous roles include being acting Associate Dean in the Faculty of Business and Management at Regent's university, and teaching German at Nottingham University. His research interests include cultural identity and cross-cultural adaptation.

“一带一路”走廊沿线的跨文化交流

The Intercultural Relations and Communication Along the One Belt One Road Corridor

The topic is about a bold vision of transforming the economic and political landscapes of Asia and Africa over next decades.

It has been widely reported in the international media as proposing the network of the infrastructure partnerships. They cross a variety of areas of energy, transportation, telephonication and IT. The transportation infrastructure is always described as the most significant one, because it strengthens the geo-political shape.

It has an enormous scope, we find 6 corridors altogether, here are Fig.1.

And as a consequence, when we link them up, we have a belt, and there is a road—it is so called as the 21st century Maritime Silk Road, which starts from the South China Sea, and across the Indian Ocean, finally it can reach the

- China-Central Asia-West Asia Economic Corridor
- New Eurasia Land Bridge Economic Corridor
- China-Mongolia-Russia Economic Corridor
- Bangladesh-China-India-Myanmar Economic Corridor
- China-Pakistan Economic Corridor CPEC
- China-Indochina Peninsula Economic Corridor

Fig. 1

Mediterranean areas. It has a number of cooperation priorities, they are:

- policy coordination
- facilities connectivity
- unimpeded trade
- financial integration
- people to people bonds

The development of infrastructure like railroads and so on has already drawn people's attention. But also, the people to people bonds, has received less attention on the media, because it's not easy to be reported in terms of impressive contracts or millions of dollars' investment, yet the ultimate goal proposed by the Chinese President Xi Jinping, is to promote the friendship among different nations of people and make a more peaceful world. In my talk, I would like to refer some of the aspects about culture and people.

Today, my talk does not focus on a material culture, I will focus more on the invisible part of the culture, which they often use the model of iceberg to explain the hidden part of the culture (Fig.2).

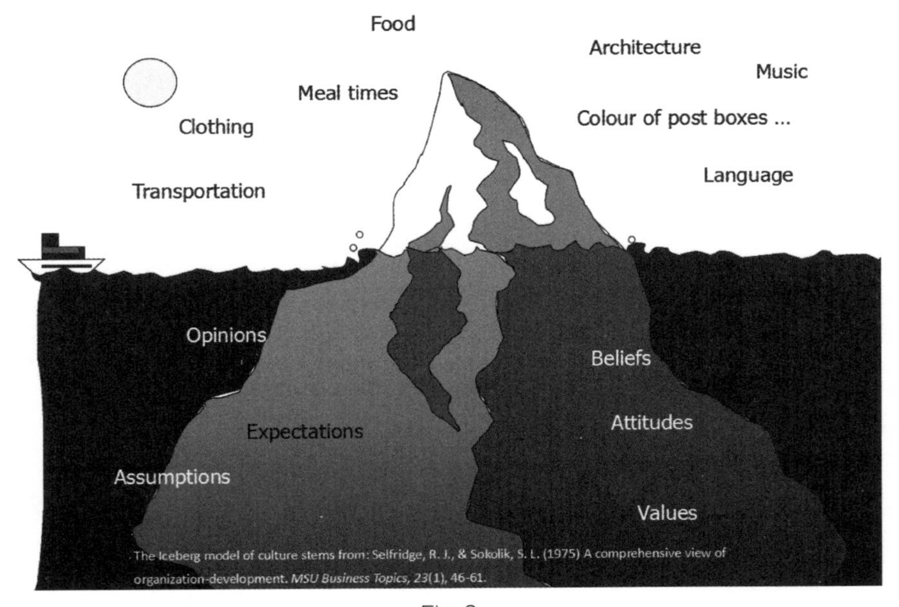

Fig. 2

As you can see, the material objects, they are above the waterline. And as we all know, the part of the iceberg, which is above the waterline, only takes up one fourth of the mass of the entire iceberg. However, the other enormous part we cannot see, which is the invisible part of the culture, the values, the beliefs, are underneath the waterline. This part is the foundation of our cultural confidence.

So, thinking about the Silk Road, to a European like me, you may have images of exhausted people among caravans and camels, travelling for the Central Asian business, bringing goods in exchange, distinguished languages have been spoken, yet such images can be romantic because of the concept of orientalism, which appeared in the book of *Edward Waefie Said* in 1978.

In his book, he examined the European politically, and he found that, deep in Europeans' thought, the most important use of Orientalism to the Europeans was for the commerce, his theory has divided the world into the East and the West (Fig.3).

It is significant that now China has close relationship with central Asian countries and become a dominant economic power in the East, and it's interesting to find that the concept of Silk Road is originated from China. Another significant aspect is that the Belt and Road concept is initially a non-western discourse. The Silk Road is also considered as a historical construct.

The Silk Road construct
• The Silk Road
• Orientalism
• Non-Western discourse
• Silk Road as a historical construct
• Allows new central Asian nations to build their heritage culture
• Soft Power

Fig. 3

On one hand, it has been constantly looking back, evoking all the connective parts to find the cultural and historical connection. And, in other purpose, it allows new central Asian nations to share their precious heritage. Here I want to address the concept of soft power, which focuses on how states and countries broaden their reputation, build their brands for their own social and cultural goods. Even though the traditional Silk Road has been lost in time, the historical relations and contemporary relations, no matter positive or negative, will be well-remembered.

So the question is that, how do the participant countries from the Corridor cooperate across so many boundaries? That will influence their intercultural relations as the One Belt One Road Program is on its way. Social psychology says that we make rational actions based on the way we collect information. It's not an objective process, but an influence of our emotions. And we found the variety of allegiances, such as political, ethnic, religious and cultural ones.

Stereotypes are very common in the cultural relations, and sometimes they can get in the way when we're having intercultural communication. So there are challenges (Fig.4).

Intercultural challenges
• Relations among and perceptions of each other in the region
• Variety of allegiances (political, ethnic, religious, cultural)
• Stereotypes
• Challenges at the societal macro level
• Challenges at the interpersonal micro level
• High degree of power distance
• High level of collectivism

At the political macro level, there are political sensitivities in the region among

Fig. 4

central Asian nations. There are also ethic differences in the same country, and perceptions that keep the balance of power have been emphasized by the people. But, it’s in the micro level where the interpersonal relationships happen. When we are having intercultural communication on the Silk Road, it’s inevitable that people from different nations will work together and communicate. The conversation may become a crash between the totally different cultures. We have to notice the high degree of power distance and high level of collectivism as well, because the most significant influence in cultural difference is the power distance (Fig.5).

Confronted with those challenges, intercultural awareness training is essential. There should be cultural exchanges in order to create an opportunity for the people who are in different cultures, to provide them with a brand-new experience of another broad and profound culture, which means cross-cultural interactions. We also need specialized training and human resource development, we need some professional workers to take the specific responsibility for the intercultural communication projects. Another significant part is, when we are communicating with others from the Corridor country, we should have the intercultural awareness, and show the respect, be transparent, and give each other enough trust. With those measures, we can make the communication much smoother. That’s all I want to share in my talk today.

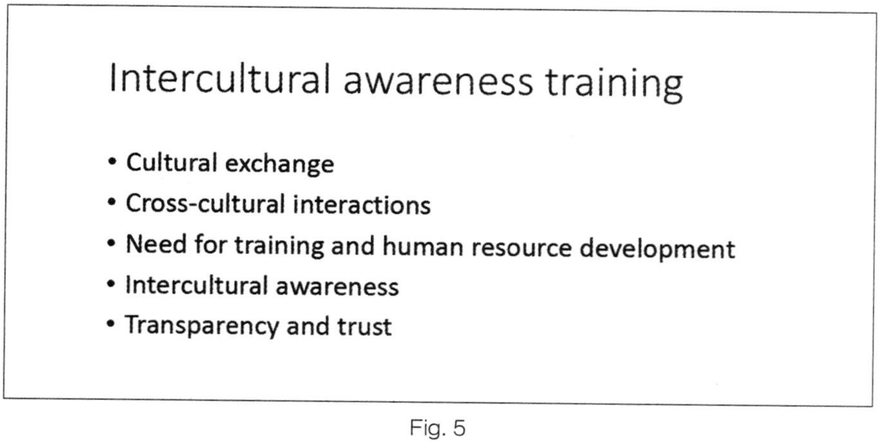

Intercultural awareness training

- Cultural exchange
- Cross-cultural interactions
- Need for training and human resource development
- Intercultural awareness
- Transparency and trust

Fig. 5

个人简介

Li Qimei

QML 文化咨询公司管理者

QML 文化咨询公司管理者、优秀的语言教师、兰卡斯特大学语言学硕士、拥有伦敦经济学院孔子学院授予的商务中文教师资格。出生于中国，在美国、日本、英国三地生活的经历使她充分理解文化差异在有效交际和相互理解中的影响力。她同爱人奥利弗·怀尔德（Oliver Wild）教授在 1992 年穿越部分丝绸之路，后于同年发表了有关他们的文章 *Old Route of Silk Road*，描述了关于丝绸之路区域的人文历史，并成为该领域内第一篇网上发表的学术文章，被学者们纷纷引用。

She is director of the QML Cultural Consultancy. A qualified and experienced language teacher, She has an MA in Language Studies from Lancaster University and Certificate in Teaching Business Chinese, awarded by the Confucius Institute at the London School of Economics. She grew up in China and lived in the United States, Japan and the UK. She have studied and worked in all three continents; these life experiences have made her aware of how cultural differences can be used to effect positive communication and mutual understanding. Li and her husband Dr. Oliver Wild had a trip along part of the Silk Road in 1992, and they wrote an article about the journey and about the history and people of the region. This was the first article in the field to be published on the web, and is now heavily cited.

穿行于丝绸之路

The Silk Road, the History

Good morning everyone, I'm honored to be here and have this talk. So, in next 20 minutes, I'll do my best to condense the Silk Road history in this talk. Hopefully it can help you gain the brief knowledge about this region.

People, in their very early history, start to trade. They want the goods they don't have, and they get what they need by exchanging goods. Trades gradually grow, from a smaller scale to a bigger scale. The significance of the Silk Road is to build a trading network crisscross European, central Asia, even Africa. Because of this kind of contact with different people, different nations, religions, ideas, cultures, technologies, even diseases spread along the Silk Road.

The Silk Road, now what we called One Belt, actually One Belt is a better description of this area, because there is no such a single route but many lines on it. Basically, in the ancient time, from the bottom of the graph, that is Xi'an, and then we go upward, passing Hexi Corridor, we get to Dunhuang, the edge of Taklamakan Desert. At this point the route divided into several branches, because of there is a large desert which people can't cross. Travelers can only choose other ways to go around the desert. In the south of Taklamakan Desert is Kunlun Mountain, and in the west of the desert is Pamir area. The north is Tianshan Mountain, and in the northeast is Gobi Desert. So the only sensible way is to go along the edge of desert, to go around of it, and stop in different points on the graph, which are oases to have a rest.

This is a clearer graph of Taklamakan Desert. It looks like that there are many rivers here, but actually most of them are dry and have no water.

The following is the photo we took in Taklamakan Desert, 1992. Miles and miles, you just see the same scenery in the desert (Fig.1).

Fig. 1

This is one of our stops in Kucha (Fig.2).

Fig. 2

I would like to talk about the early history of the region. The western side of the trade route appeared to have developed earlier than the east principally because of the development of the empires and the easier terrain of Persia and Syria.

This is the Persian Empire around 500 B.C.E., as you can see. Because of the relative stability in the region, the terrain is much easier to control, the people are not so hostile like people in Taklamakan, so the trade developed much easier, and it's more convenient for caravans to go through the western area.

After the Persian Empire, in about 330 B.C.E., Alexander the Great conquered this area. Although Alexander the Great's army only stayed in the area for five years, it has a long-lasting influence. Because wherever Alexander's army arrived, they left some soldiers there to supervise and guard the area, so the soldiers lived and married the local people.

Therefore, the soldiers brought their language and culture to the conquered area. Later on, their language and culture mixed up with the native culture, which formed what we called Hellenic Culture.

Then this area has separated again. In the northern part of the region, a group of people called Sogdians, they were very skillful in trading. To learn more about the Sogdians, you can visit the UNESCO website of the Silk Road.

In Eastern Han dynasty, about 25–220 A.D., here's a Chinese ceramic statuette of a Sogdian caravan leader of the Silk Road, shows what Sogdians like in ancient times. This one is from Tang dynasty.

Down slightly to the south region, many tribes ruled this area. Around 100 B.C., Yuezhi people came, they were driven out from their original homeland by Xiongnu, so they came here and settled here, adopted Buddhism as their own faith, and incorporated the local culture, and then gradually became what we called Gandhara Culture.

Let's have a look at the eastern end of the Silk Road. In 221 B.C. Qin Shi

Huang ended the Warring States and united China. His main achievements involved:

- unification of the language
- unification of the measurement
- location of the capital Chang’an

In Han dynasty, we had the first prosperous period of the Silk Road. One of the most significant achievements is Zhang Qian’s Journey to the west. Around 130 B.C.E., because the northern border of Han dynasty has constantly been invaded by Xiongnu, Zhang Qian with his team of around 100 people, tried to link up with Yuezhi, to defend Xiongnu together. He didn’t come back until 125 B.C.E. because he was captured by Xiongnu people on his way back, but when he arrived his home country, he brought back a lot of information about the unknown countries on his route.

Next part, the name of the Silk Road. There was no specific name for the route that linking the west to the east. In 19th century, a German scholar called Ferdinand von Richthofen initially described these routes as the Silk Road because the silk was the most demanded commodity for the Romans.

The most significant commodity along the Silk Road is not silk but religion. Buddhism came to China along the northern from Indian. We still have some grottos in the northwest of China now. These grottos are valuable source or information about the Silk Road. Along with images of Buddhas and Bodhisattvas, there are the everyday life of the people at that time, through the images of their celebrations and dancing, we can have an insight of their customs and costume.

The height of the Silk Road was in the Tang dynasty. After the Han dynasty, several kingdoms came out, and Tang dynasty was another unification, which was relatively stable in politics. In this period, in the 7th century, the Chinese Buddhist monk Xuanzang crossed the country to Indian to obtain Buddhist scriptures. He carefully recorded the cultures and styles of Buddhism along the

way. On his return to the Tang capital Chang'an, he was permitted to build the 'Great Goose Pagoda' to house more than 600 scriptures he brought back. His travels were dramatized in the popular classic *Tales of a Journey to the West*. Chang'an, now we called Xi'an is one of the largest cosmopolitan cities in the world, by 742 A.D. The population was almost 2 million, with as many as 5000 foreigners, including Turks, Iranians, and others from along the Road, as well as Japanese, Koreans, Malays.

After the Tang dynasty, China entered to another instability with wars, and the silk road subsided gradually. Since then, about 700 A.D, Islamic was about to rise. The trade relations soon resumed, but the Moslems started to play the middleman. Also the sea route was explored at this time, the Sea Silk Road eventually was formed, and became more profitable later on. But the final shake-up didn't occur in the country, but occurred from different directions because of the expansion of Mongols.

In summary, from its birth before Christ, through the height Tang dynasty, until its slow demise six or seven hundred years ago, the Silk Road had played a significant role in foreign trade and political relations. It has left its mark on the development of civilization on both side of the continent.

Now, with the renewed interest, the edge of Taklamakan see the international trade once again, on a scale considerably greater than the old past.

个人简介

Stuart Walker

英国可持续设计的先锋学者，英国兰卡斯特大学想象力研究中心创始人和管理者，英国肯辛顿大学客座教授，加拿大卡尔加里大学名誉教授

英国可持续设计的先锋学者、英国兰卡斯特大学想象力研究中心创始人和管理者、可持续设计中心主席、英国肯辛顿大学可持续设计客座教授、加拿大卡尔加里大学名誉教授。他的研究以实践为基础，探索了可持续发展中环境、社会和精神等方面的重要性。他在加拿大和英国获得很多奖项，他的论文已经在国际上出版发行。他的概念设计作品已经在加拿大、澳大利亚、意大利、伦敦设计博物馆展出，最近又在英国湖畔区约翰·拉斯金的住处——布兰特伍德进行展示。已经出版的专著包括《通过设计实现可持续》《设计精神》《可持续设计手册》(合著)《设计可持续性》等。即将完成的专著包括《生活设计》和《设计根源》。

Stuart Walker is Chair of Design for Sustainability and a Founder-Director of the Imagination Lancaster Research Centre at Lancaster University, UK. He is also Visiting Professor of Sustainable Design at Kingston University, UK, and Emeritus Professor, University of Calgary, Canada. His distinctive practice-based research explores the environmental, social and spiritual aspects of sustainability. He has received numerous awards in Canada and the UK and his papers have been published and presented internationally. His conceptual design work has been exhibited in Canada, Australia, Italy, at the Design Museum, London and most recently at Brantwood, John Ruskin's house in the English Lake District. His books include: *Sustainable by Design*; *The Spirit of Design*; *The Handbook of Design for Sustainability* (ed. with J. Giard), and *Designing Sustainability*. Forthcoming books include *Design for Life* (Routledge, May 2017) and *Design Roots* (Bloomsbury, June 2017).

为生活设计：在抽象世界中创造意义

Design for Life: Creating Meaning in a Distracted World

Good morning, everyone. It's wonderful to be here and thank you for the invitation. As my self-introduction, I'm the professor, or designer of sustainability, mainly product design, not fashion design. So what I focus on is the object, the factory products, and design for sustainability. And I hope what I'm going to say will be applicable across the broad-area design, not just object design. So, in this presentation, I'm going to talk about my research approach in which I looked at the relationship between human values, sustainability and the forms of tradition and localization.

So, I'm going to begin by contrasting Modernity, it's about innovation, change and consumerism, which is well-known as very under-sustainable, with all traditional practices saying that many ways can form the contemporary ideas about sustainability.

I'll talk about what how we might learn from these traditions, and we might develop a different approach to design. And I'll finish by showing you some examples of my own design work.

This is what we do to the natural environment. We dig up the earth. We pollute the air and the oceans, and we have been doing these since the Industrial Revolution. This is modernity, which is rooted in scientific advancement, technology and consumerism. To drive a consumption-based economy, we have to control the Earth. But instead we talk of it, in terms of resources, for economic growth. In effect, we have an instrumentalized planet, I come to see it, and something to be exploited, for monetary gain. And we call this, "progress". In the name of it we consume many things, that we don't actually need, and we create wastes through plastic packaging and short-live products, we create a world of trivial novelty and endless choice. It's exceedingly harmful, they also represent a crisis of values. It's reducing the quality of the environment, and our own quality of life. And it's compromising the ability of the earth to sustain healthy forms of life. Perhaps, more than anything, modernity represents a loss of meaning in our lives, this is the so-called a disenchanted world view. We are all consumers, always waiting for the next thing, the next mobile phone upgrade, the next broadcasted movie, the next fashion trend. But we're not always living this way, this is a product of modern era. Before this, we have lived a sustainable balance with the Earth. When we are in the time of Modernity, we have become materially rich, but spiritually impoverished.

Pre-Modernity tended to be spiritually rich, but there is no doubt also, that compared to today, there was material hardship. Somehow, we have to find the old balance we have formed. We'll have fewer material expectations by reducing our exploitation of the Earth. And in the process, we find ways of living that are spiritually richer and more fulfilling. Pre-modern, more traditional ways of living are generally characterized by rather different values, and the quite different world view. Let's take a look at how this different sensibility affects the nature of human actions, and the way people have for centuries to satisfy their material needs.

This is the example of the West Coast of Canada. Here the first nation's people carried out sustainable forestry practices. Here we can see, the bark is taken off from one side of the tree only, and it will be used to weaving baskets. The practice doesn't kill the tree, but in time the tree will heal itself. Very similar practice was implemented by the indigenous people in Australia. Here we see, red gum trees on the banks of the Murray River, their bark will be removed for canoe making. Traditional ways of thinking and doing are considerably

different from modern sensibility. They tend to involve the sense of beauty and responsibility, not just to others in their community, but also for the teaching of knowledge and wisdom. This sensibility nourishes the common sense of identity, culture and continuity, and it enhances the sense of spiritual well-being, and sense of purpose and significance (Fig.1)

In Victor de Sousa's *UMA LULIK*, the documentary about the traditional sacred house, the villager says, "We didn't come up with the wish to build this house. It is a wish preserved from ancestors" (Fig.2).

Fig. 1

Fig. 2

The woman in this picture, she is practicing the ancient craft of harvesting and weaving. She's dealing with this golden-colored fabric. She was taught by her grandmother, who learned it from her mother. But she says that her daughter will have to continue this tradition so humankind can benefit from it (Fig.3).

And in England, very close to my own home, a shepherd expresses the same statement. "Some people's lives are entirely their own creation. Mine isn't. The flocks remain, the people change over time. Someday I will pass them on to someone else" (Fig.4).

In these examples, we have seen that, all around the world, such practices are not aware of individual choice. There is a sense that one has to continue the tradition from the old people, that is a duty to one's culture, one's community, and one's own sense of identity. This concept is different from individualism: put emphases on personal choice, which we find in modernity. Our challenge today is to develop new creative ways or the combination of the sense of belonging responsibility and identity we find in traditional cultures, with the capabilities and benefits that have been developed in modern times. We'll go on to go back to pre-modern times, where we can learn from the traditional practices to develop a new outlook that brings together traditional and modern perspectives. In terms of the making opportunity of culture, they passed on the

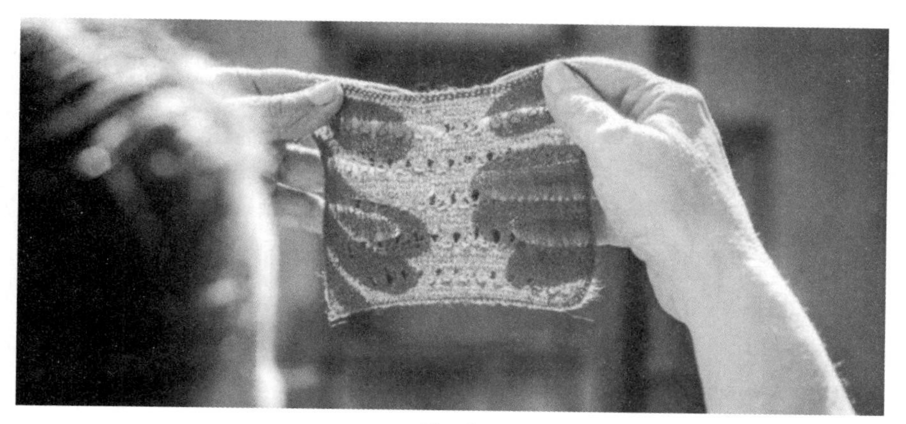

Fig. 3

Fig. 4

traditional methods from one generation to the next. These methods will be renewed over time, but during its change or development, it tended to be slow. People are suspicious about the innovation because if things went wrong, they could find themselves in awful danger, they would be lack of food or shelter. So, they would like continuity rather than change, and while there was less choice, they would have a great sense of dependent on community, and have deep knowledge of local conditions, which they had the best interest to look after, because they survived on the local knowledge.

After the Industrial Revolution, we are in modern times. And we have world-marketing economy, consumerism and globalization, but also, unsustainability, because of the consumption and advanced science and technology. We might expect modern approaches to prevent environmental crisis, this is called eco-modernism, and it's coming smoothly into our current system. Companies and politicians like it more, however, the potential to contribute significant and long-lasting change is limited, it can be counter-productive with the old same system of consumption growth and waste.

So if we want to tackle the problem of sustainability, we have to develop a different outlook. We have to develop a way forward the old system from the traditional ways of knowing. In attempt to integrate the benefits of pre-modern and modern times, we develop a new world-view, which I called after-modern, it's rather different from post-modern.

Traditional philosophical spiritual teachings from Plato and the Bible in the West and Confucius' thought in the East, always talk about human values. They are concerned with how we should live in the world, how we should regard material possessions, how we should behave towards one another, how we should regard the Earth, and on the personal level, how we should live, if you are finding some degree of happiness and fulfillment. These thoughts are very ancient and their meanings are motivated, we cannot apply these methods into inductive reasoning.

Heidegger distinguished the thoughts into scientific knowing, which concerned prediction and control, and social ways of knowing, which is concerned with values, beliefs, and experiences. When we talk account of human experience, it does not refer simply to the experience of one individual, but an accumulating heritage of the human experience. Recognize in the contribution by those who came before us, it's fundamental to tradition, and it's embedded in the spiritual teaching (Fig.5, Fig. 6).

Rather than implying the methods of science, more appropriate approaches to comprehend the ancient thoughts is the interpretive method. When dealing with texts, especially traditional spiritual texts, the interpretive method is used

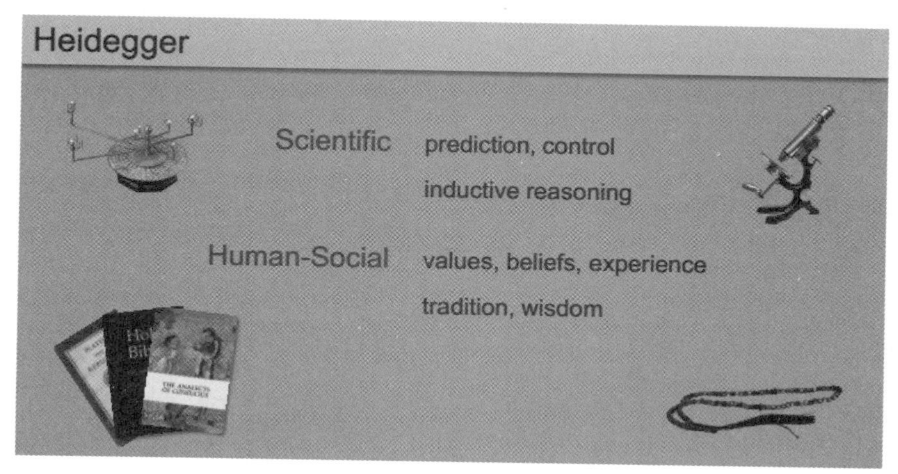
Heidegger
Scientific
prediction, control
inductive reasoning
Human-Social
values, beliefs, experience
tradition, wisdom

Fig. 5

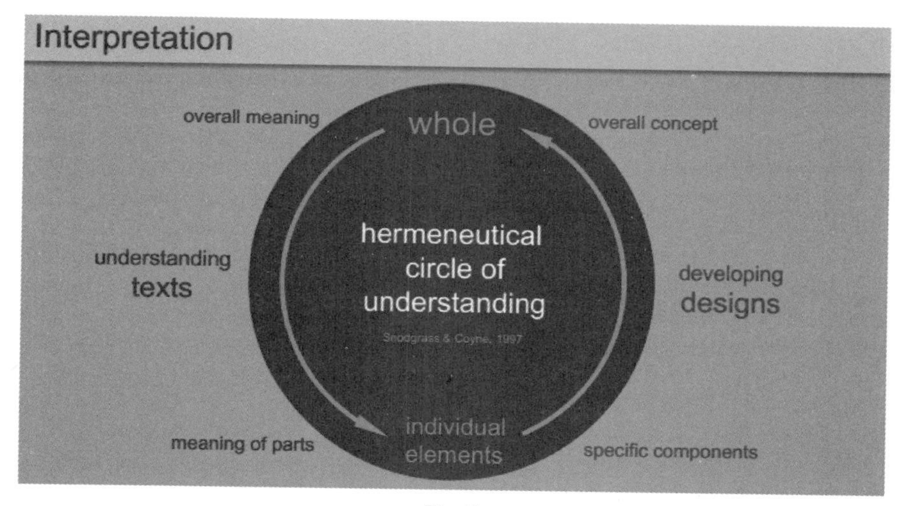
Interpretation
overall meaning
whole
overall concept
understanding
texts
hermeneutical
circle of
understanding
developing
designs
meaning of parts
individual
elements
specific components

Fig. 6

as hermeneutical circle of understanding, the rotates between the hole and individual parts. In the case of a text, we cannot understand the meaning of the part unless we understand the whole, and we cannot understand the whole until we understand the individual parts. A similar situation is encountered in the process of designing, or any creative activity. We consider the whole and the individual parts and keep moving between the two. There are generally four levels of interpretation that is quite traditional in teachings. Let's go from the Literal, through the Moral, and the Allegorical, to the Anagogical, or the spiritual meaning. As we are moving through these, we can get close to a deeper understanding of all the great tradition.

So here, we find the anagogical level, the spiritual level, whose values transcend social cultural differences. These are known as self-transcendence values for yourself. They are concerned with the things like welfare for others, stewardship and natural environment, respect for the tradition, customs, including the spiritual or religious practices. But this heritage has been eroded during the process of Modernity: in the 18th and 19th centuries, with their pragmatic individualism degraded and corrupted the psychological heritage of axial man. The post-war growth in consumerism has further degraded this heritage. And the recent rise in personal digital device is affecting us as well.

Modernity, has emphasized philosophical materialism, individualism and consumerism, which is built on the market-based system that full of greed, dissatisfaction and fear. And it tended to foster the self-enhancement or self-regarding values, such as social status, achievement and personal success, material wealth and the pursuit of pleasure and power. Significantly, these are all oriented towards extrinsic or external goals and rewards.

But today, in the self-transcendence values, we focus more on traditional teachings, practices, localization and sustainability. They are all intrinsic goals, or internal goals or rewards, such as seeking excellence and finding satisfaction in one's world. Through the interpretation, we may have further understanding of traditional teachings and practices, and make it relevant to the present. The ability of designing is also interpretive. Self-transcendence values can link to

cultural relevance from conventional design practice.

Praxis is the action based on reflection that changes the situation for the better. Conventional design practice is a means to primarily commercial ends. In contrast, praxis is concerned with the wise determination of ends, and it's the means of attaining those ends. It's been proved in scientific approaches because it integrates actions with values and beliefs. Praxis also consciously resists or interrupts the "*hegemonic status quo*". In conclusion, we can say that Progressive Design Praxis is a form of design practice that aims to change the situation for the better by striving to interpret, understand and apply the ethical values and notions of virtue found in the philosophical and spiritual traditions of one's own culture. I'd like to introduce the progressive here, is to make progress gradually in the present condition.

Progressive Design Praxis transcends empiricist theories of knowledge, it dominated the contemporary thoughts in a period, and still dominates today. It challenges the prominence of theory, recognizes that the reason cannot solve contradiction, and raises the significance of practice.

To translate this kind of praxis into design actions, we can do this from the practical level. To provide functional benefits, or also take into consideration of the environment-friendly designs. We can consider the social meaning, and we can also concern the personal meaning, in terms of spiritually well-being, or personal values. We also have to consider the economic means. Economic is the mean to achieve the other three, and it is not an end of itself. So, these are the Quadruple Bottom Line of Design, we can show the relationships between them like this: these factors are not in competition; they all need to be addressed together. If we only have a look at the practical meanings of design, we will see a narrow world. If we also focus on the social meaning, then we have a wider perspective, which will make us more influential in the late-modernity period today. But if we focus on personal and spiritual meanings, our design activities will be far more comprehensive. When we're implementing the Quadruple Bottom Lines of Design, we're actually building expertise over our own lifetime. So, we start with the practical level, on the bottom left of the diagram, but as we move through time, we also move upwards, go higher with our inner consideration. We start to reflect what we are doing and turn it into self-transcendence values, then we start the critique of consumerism, and then we start to critique the worldview. And we begin to change our priorities, so we make different design decisions (Fig.7).

I've been doing this kind of work, exploring these ideas over many years and It'll be finished by showing you my explorations which I attempted to embeded the values, sustainability and nature into our material culture. So, let's have a look at some of these objects. They're not commercial objects, but practice-based design research. They start off at the practical level, making incremental changes. I started off at the practical level many years ago, making this kind of design, and reuse the material (Fig.8).

We often think of design of sustainability as being a long-lasting product, so this is the exploration of one series which looks for ephemerality, objects that come together for a while to make a function then it becomes distributed back into the environment after they're used.

So, moving up a level to inner reflection and questioning the current condition, there is a design for a mobile phone few years ago, it's an inconvenient mobile

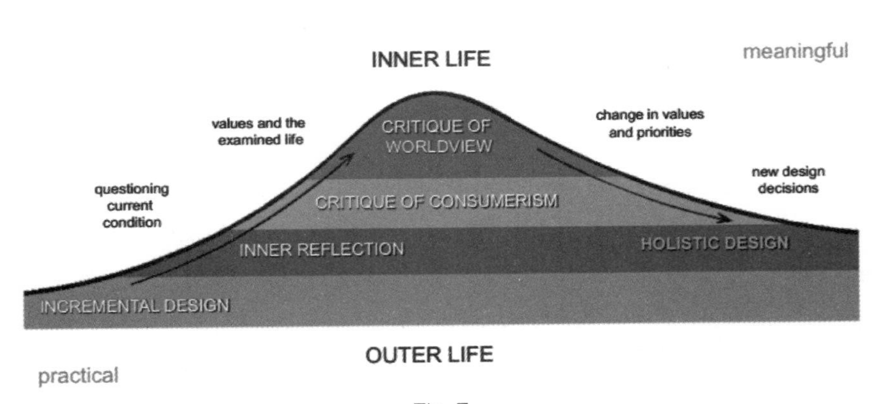

Fig. 7

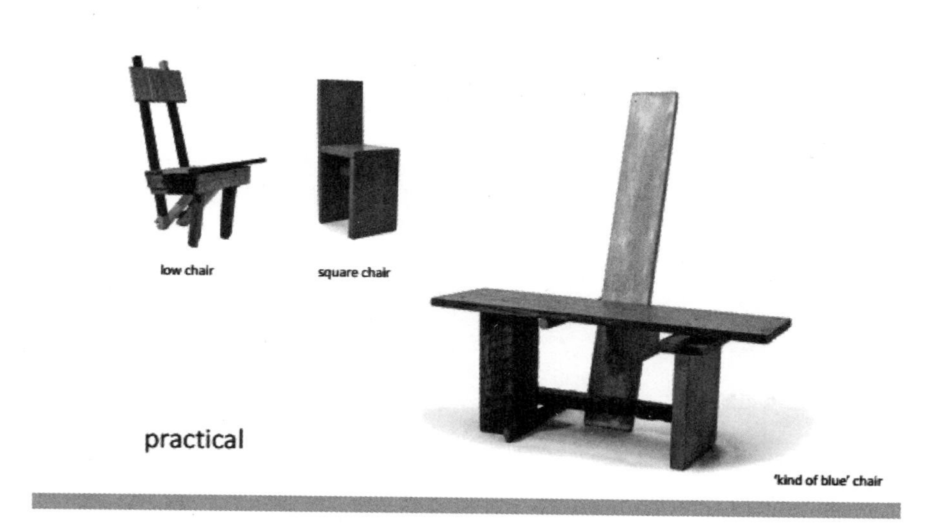

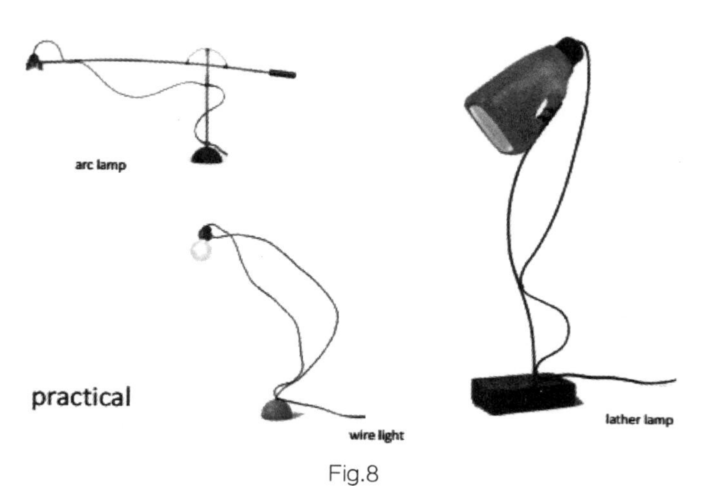

Fig.8

phone, it's reasonable for its inconvenience which is about addictive use. Inside the package, there are a bunch of parts, if you want to use it as a phone, you have to assemble the parts, and you pay attention to your phone calls and messages, and then you take it apart and put it back, so that it won't interrupt you when you are walking down the streets or in conversation with someone else. It won't distract you from where you are, then you have single point of attention when you have the phone call, and then put it away. And it has many variations on that, from the most inconvenient, the most upgradable, and the most sustainable aspects, there's various other are options that are the most convenient but also the least sustainable (Fig.9).

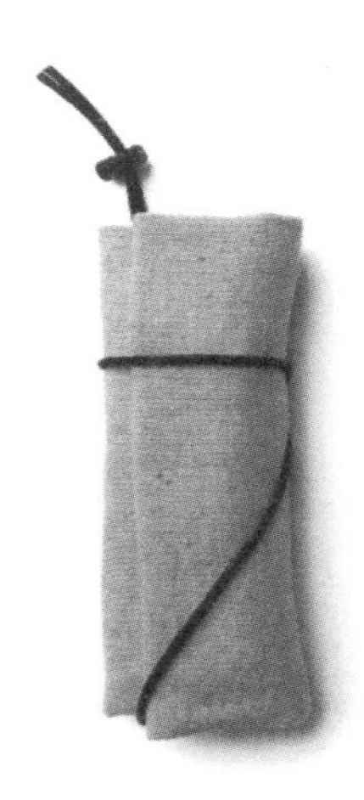

Wrapped Attention
mobile phone

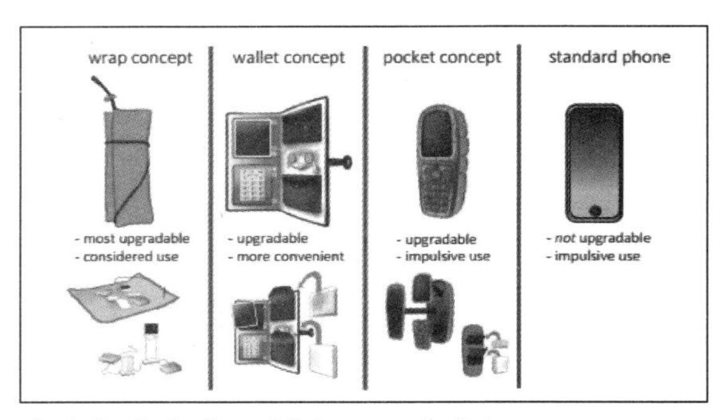

Production, Service, Economic Factors
- permanence – it evolves over time
- ≈ 80% less waste
- product-service system
- local +mass production; local servicing

Use Factors
- considered use – challenges 'convenience'
- user in charge of technology
- a phone for life – maintain and update
- offers *meaningful* choice

Fig. 9

This is an object which basically, just literally, ties together the technology with the natural environment to create an object that combines the two to remind ourselves that these psychological objects come from the natural environment, and when they have been used, and go back to the natural environment (Fig.10).

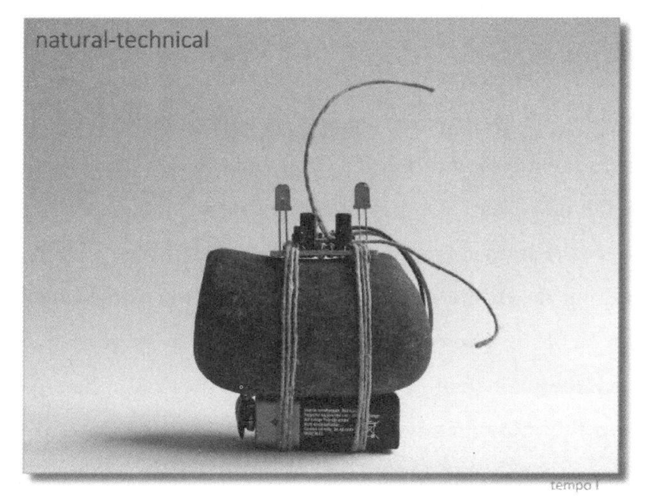

Fig. 10

Then we move up to another range, the critique of consumerism. To create a 100% sustainable product, we should focus on the issue that modernity is basically ignored, which is spirituality. Here are prayer stones, they focus not on utility, but spirituality. Here's another one. This is an object for trans-religious. This object means that there is some little light coming through the narrow gate (Fig.11).

Moving up again to our critique of the worldview. These are visual statement using the same kind of techniques of object design but to make visual statement of our current condition.

prayer stones - a trans-religious, spiritually useful object

Fig. 11

This one is called “last supper”. I take this idea literally because we’re consuming the Earth (Fig.12).

This is called “beyond words”, the transition to modernity with the emphasis of word, and the emphasis of intellectual knowing. It causes intuitive apprehension and the spiritual way of inner knowing. It is explicit knowledge, descriptive knowledge, and intellectual knowledge (Fig.13).

This is about the digital culture, digital slavery of your life, being shackled, wherever we are climbing the mountain, walking on the beach, we are never really there because we are connected to “there”, like iPhone and other similar devices (Fig.14).

This is called Oedipus eyeglasses. Oedipus is about an existential crime by killing his father and marrying his mother and having children with her. This is called an existential crime because it can never be repaired and you can never make good again. And the punishment for the existential crime is self-blindness. And we today are creating an existential crime, we are destroying the Earth, and the punishment is our own self-blindness in doing that (Fig.15).

This piece is called neophobia, the commodify cures for the illness of consumerism. About the consumerism, whatever the issue, it can find a way of selling something to you. Even the problem that the consumerism is creating, it can make a product and sell it back to you. In the jars, we can see

Fig. 12　　Fig. 13

different senses, so this is the cure we buy back to cure the problems created by the consumerism, like William Davies said, “Our anxiety is their revenue opportunity”(Fig.16).

This is water of modern history. At first glance, the water looks beautiful. It looks like spring water, rain water, mineral water, well water, island water, and canyon water. But if you see the small letters on the top, you can see that they’re silent spring, acid rain, conflict mineral, fracking well, three-mile island and torrey canyon, all the environmental disasters. So the modern history of water has another kind of commodification, in these bottles, and contamination (Fig.17).

So moving up again, the critique of the worldview. I have been done this journey, we then come back to objects of use, objects of utility, we’ve been through that journey (Fig.18).

Fig. 14

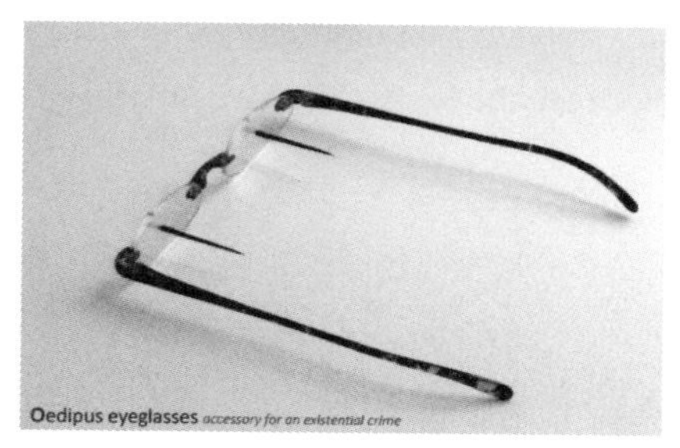

Fig. 15

Fig. 16

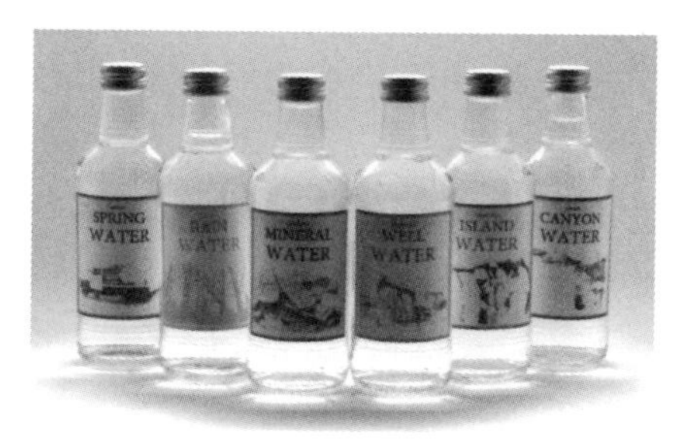

Fig. 17

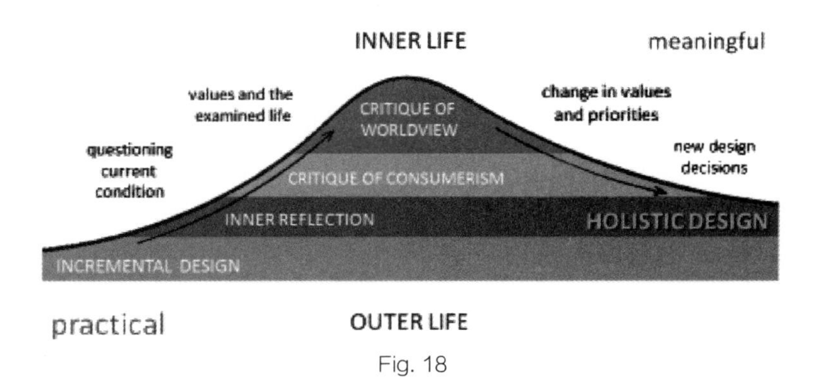

Fig. 18

Here is a floor lamp, which combines the technical with the natural. Those components can be very easily separated, which go back to the natural environment without any other effects, and the technology can be reused easily, because the technology wasn't designed for this particular lamp (Fig.19).

Fig. 19

This is the desktop version for the lamp, using the same idea (Fig.20).

This is chess set. Only in left we see the traditional, pre-modern chess sets in 1849, which we can see the king, the queen and other images. In this version of 1923, the modernist Bauhaus made this chess set without the concept of hierarchy. It's a very wonderful example. In this Balanis chess set, I tried to combine both the traditional design and this modern design (Fig.21).

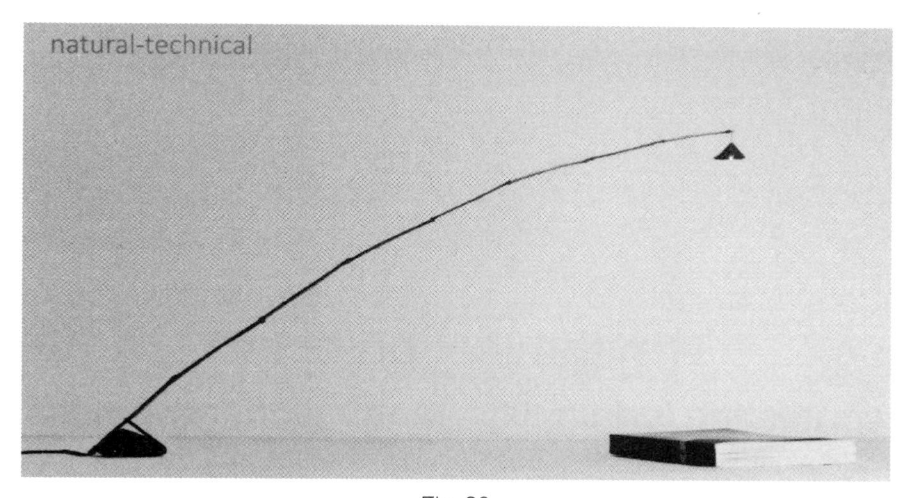

Fig. 20

traditional design
Staunton chess set
1849

modernist (Bauhaus) design
Hartwig chess set
1923/24

Balanis chess set

holm oak and flax

Fig. 21

Fig.22 is called rhythmical objects. In this kind of objects, they are making the process, using the process, and there are relations to spiritual activity.

So in the making of these objects, and in using these objects, there's a rhythmical, physical activity, which is conducive to meditation and contemplation. Here are some examples (Fig.23).

The final concept is called seeds of change, which allows us to see what is right before our eyes in a new way. You will realize too late that wealth is not in banking accounts. Thank you for your attention!

rhythmical objects

contesting 'originality', innovation, newsness

making linked to well-being, nature, locale and tradition

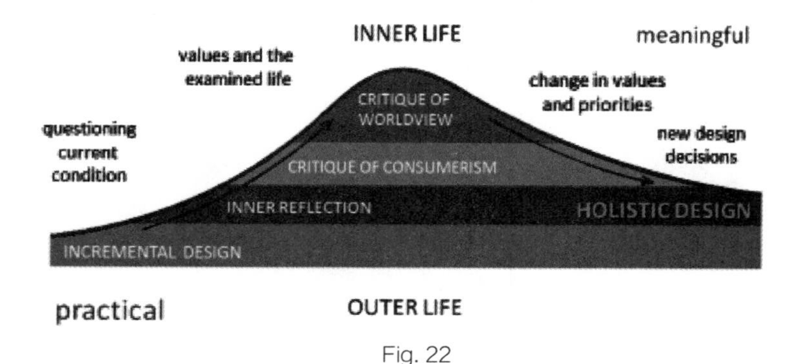

Fig. 22

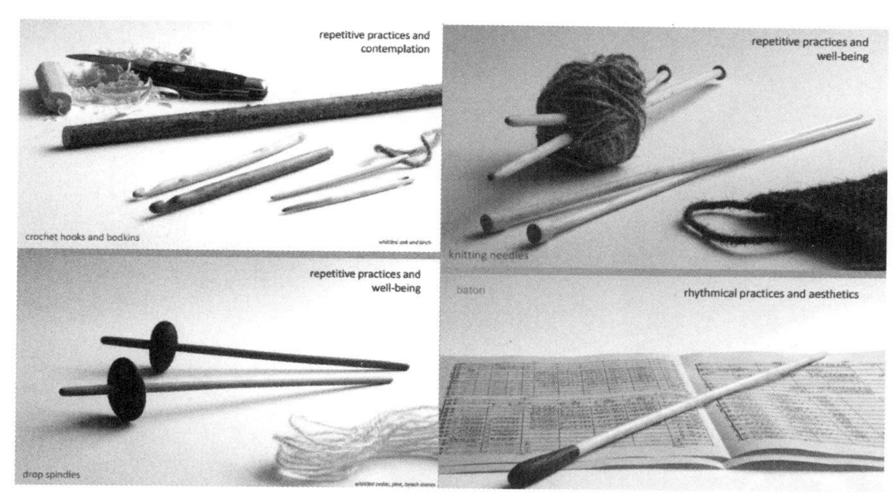
repetitive practices and contemplation
crochet hooks and bodkins
repetitive practices and well-being
knitting needles
repetitive practices and well-being
drop spindles
baton
rhythmical practices and aesthetics

Fig. 23

个人简介

Sergey A. Yatsenko

俄罗斯国立人文大学历史和文化理论学院教授

1957年出生在苏联罗斯托夫市，现俄罗斯国立人文大学历史和文化理论学院教授，自1975年起参与东欧和中亚考古考察。自2001年起为欧洲考古学家协会（EAA）会员，2011年起为国际社会标记研究会员。2002年于俄罗斯国立人文大学获得博士学位，博士论文课题为“习古伊朗语民族和历史文化重建方法”（2002年12月，俄罗斯国立人文大学）。

谢尔盖·A.亚岑科教授在考古、艺术和宗教领域发表了250篇著作，内容包括伊朗和土耳其民族的服饰、艺术肖像学、游牧民族社会和种族历史等。

Sergey A. Yatsenko was born in 1957.05.15 in Rostov-on-Don city (Soviet Union/Russia). He has taken part in archaeological expeditions' participation since 1975 in the Eastern Europe and the Central Asia, and owned the membership in European Association of Archeologists (EAA) since 2001, in International Society for Mark Studies (Signum), since 2011. His Post-Doctoral Dissertation was "Costume of the Ancient Iranian-Speaking Peoples and the Methods of its Historic-Cultural Reconstruction" (December 2002, Russian State University for the Humanities). He is the Professor of the Russian State University for the Humanities/RSUH (Moscow), and is deep of the History and Theory of Culture (since 2005). Vice-dean of the RSUH Fac. of the History of Art in 2001—2010.

He is the author of about 250 publications (mainly in Russian and in English) on archeology, art religion of the ancient Iranian-speaking and Turkic-speaking peoples (mainly nomadic) — on their costume, art iconography, burial traditions , nomadic groups social, ethnic history, nishan/tamga clan/family signs.

5–8 世纪丝绸之路服饰研究的些许问题

Some Problems of the Silk Road Costume Studies for the 5th–8th Century

Any good fashion designer or couturier has to properly study the costume of the past belonging to both his or her own land and many other countries; otherwise it is impossible to intuitively feel important mechanisms of change happening in people's looks. Those designer's ideas which are considered "unexpected and modern" by fashion-impaired people often come from the past. I have been studying the history of the ancient costume for forty-two years so far. In 1987, following the recommendation of the well-known specialist in ancient cultures of Central Asia academician Boris A. Litvinsky (Yatsenko 2000), I got interested in bright costume complexes of peoples who lived in western territories of modern China in oases of Xinjiang, and in northern parts—in the lands of nomads. The cultural influence of the both (the northern nomadic neighbours and the Han people 汉人) was felt in oases. At the same time, the bright and specific costume was being formed in each oasis. A great number of imported fabrics, different ornament and plot combinations were used in oases due to imported trade routes hosted there. It was relevant for the local population to preserve and develop their own costume as a visual symbol of their ethnic identity under the condition when both the Chinese Empire and nomads were trying to extend their authority. All above said makes the study of Xinjiang Oases materials relevant to a historian. Today, the costume of the first millennium of the Christian era is by far the most detailed for Khotan (和田，于阗) (it is amply documented, though uneven in time, and we can determine the nature of its evolution) (Fig. 1) and Kucha (龟兹，where we have detailed information for the short period of the 6th–8th cc only, Fig. 2). The sources of information for the two oases are quite different. For Khotan they are terracotta and authentic burial clothes, terms in some texts. The costume of Kucha is amply documented in Buddhist cave paintings and presents a great variety of forms.

Fig. 1 Costume complex Costume of Khotan Kingdom of the 1st–8th cc. CE (1-2—after Yatsenko 2000; 3-4—after Fabulous Creatures 2001; 5—after Gropp 1974; 6—after Li Xiaobin 1995)

Why is the costume of the two oases above all so interesting for us? I believe the reason is as follows: it had an impact on the costume of other often far away countries, thus, Xinjiang cannot be viewed as a peripheral region in the world history of the costume. It happened mainly due to the political influence and wide spread of the Turks (突厥)on the territory of Eurasia.

First, even in early terracotta from Khotan of the 2nd–4th cc. CE, we can see the long sleeved coat with two lapels which started to spread beginning from the 7th cc. CE, and nowadays it dominates in the formal and gala costume of the whole world in the form of the European jacket (Fig. 3).

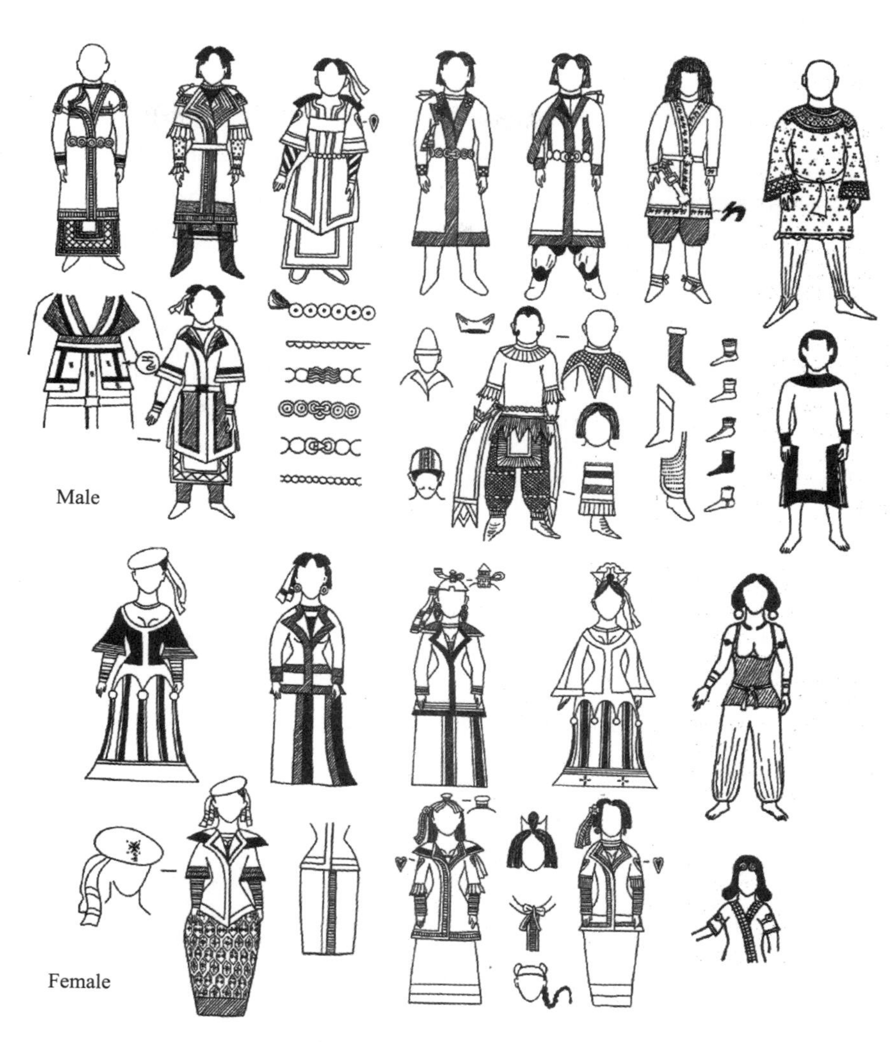

Fig. 2 Costume complex of Kucha peoples of the 6th–8th cc. CE (after Grunwedel 1912; Le Coq 1977)

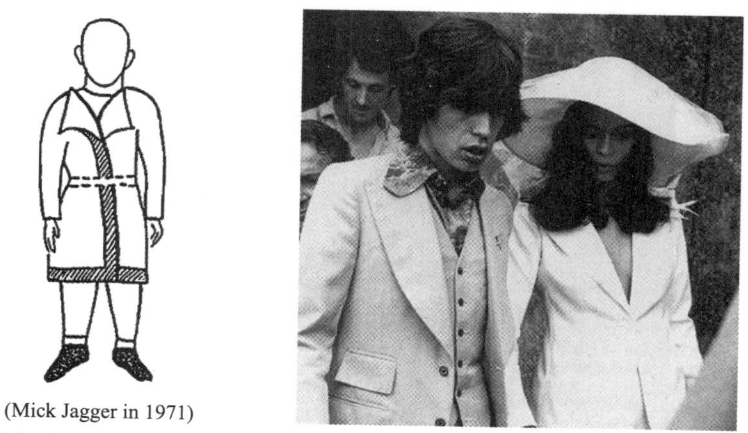

Fig. 3 The long sleeved coat of Khotan of the 2nd–4th cc. CE with two lapels the modern jacket

Second, Kucha may be considered the homeland of the low rectangular neckline which is known for short upper garments in the costumes of Eurasia. In Kucha of the 6th–8th cc, low neckline was worn by both aristocratic women and common dancers, the décolletage could also be hemicircular [Fig. 4 (1)]. Another type of neckline we know at that time for Khwarezm [Fig. 4 (2)]. The type of Kucha pelerines was seen later for the Turkic Uygurs of the same region [Fig. 4 (3)]. Probably, it did not only generate men's interest but also had ritual functions (for example, such neckline was depicted in Khwarezm and for Uygurs for women who mourned the dead).

Third, for Kucha women we see a very slim waist which, probably, was a result of corset usage, likely beginning from adolescence [Fig. 5 (1)]. More rare it was used in Khotan (Taryshlak) [Fig. 5 (2)]. Probably, it was the earliest region of corset usage. It could be worn beginning from adolescent age as in Kumtura we can see the depiction of a local ruler's daughter—a girl with a very slim waist.

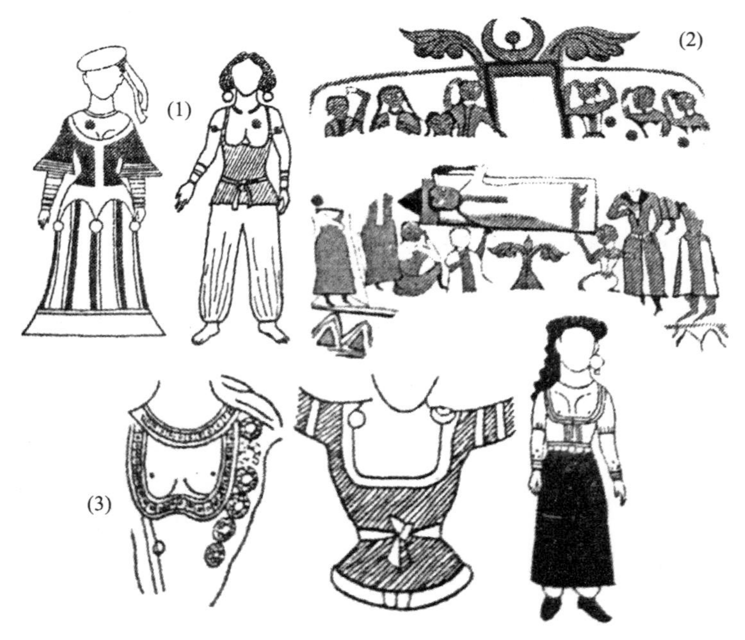

Fig. 4 The low neckline in Kucha (1), in Khwarezm (Tok-kala), (2) in the western Khotano-Saka regions (Toqquz-Sarai), (3) and in Uigur Turfan kingdom (4).

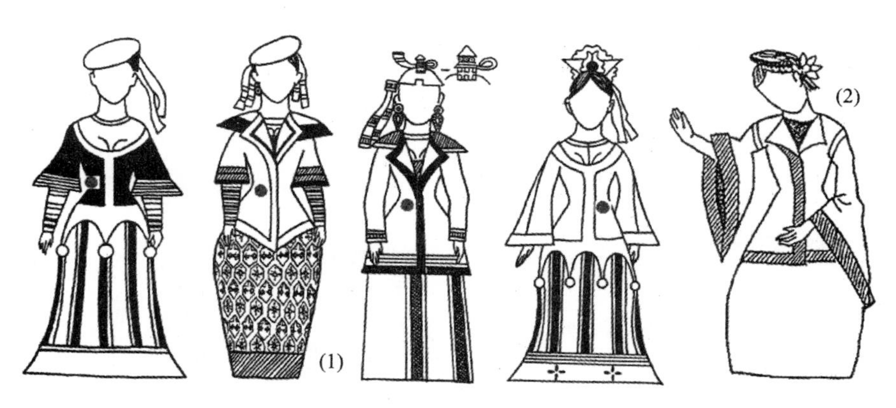

Fig. 5 Corset usage in Kucha (1) and Khotan (2)

Fourth, for the Turks of the late 6th–early 8th cc we unexpectedly notice some influence of the costume of Kucha Oasis (the latter being under their political power for some time) (Yatsenko 2009, figs. 15, 17, 19). For example, on the sarcophagus of Yu Hong, the Sogdian 'sabao' from Tayuan city, who died in 592 we see only one Turkic personage presented in panel No. 4 [Fig. 6 (1)]. The costume of this Turk is very unusual. A short upper garment has short sleeves flared at their end, and the undergarment is longer and has usual narrow sleeves. The edges of the cuff are scalloped. Such specific combination of two garments worn on the upper part of the body was typical for the population of Kucha. Another original costume element here is a bunch of five short ribbons hanging from each shoulder and typical for noble men exclusively in Kucha [officers usually worn them only on one shoulder, Fig. 6 (2)]. The decorative breast ribbon, the trousers and shoes types are also characteristic features of the Kucha costume.

In the some burial terracotta Mingqi（冥器）of the Early Tang time with the realistic Early Turks the long sleeved cost was green [Fig. 6 (3)]. According to Xuanzang who watched in 630 the Western Turkic Kaghan Ton-yabgu with his escort in an everyday situation during hunting and specially marked the exclusiveness of a green caftan for him only. Such green clothes were popular in China during the Tang dynasty; later in the Song time, clothes of green and bright red colors were perceived as the most unbearable manifestation of the barbarian trend in the Chinese costume at the time which had spread since the epoch of the Northern Qi (550—577. A. D.). But the special popularity of green for men’s upper garments for Turks in the 7th cc., we observe in the abovementioned Kucha Oasis which bordered on the territories where nomadic Turks roamed and camped [Fig. 6 (4)].

Fig. 6 Influence of the costume of Kucha Oasis for the Early Turks: (1) Turkic man with Kucha’ costume elements on sarcophagus of Yu Hong, the Sogdian ‘sabao’ from Tayuan city, died in 592, panel No. 4; (2) Kucha short upper garment with short sleeves flared at their end, and the undergarment is longer and has usual narrow sleeves; male bunch of five short ribbons hanging from each shoulder; (3) burial terracotta Mingqi of the Early Tang time with the realistic Early Turks, the long sleeved coat is green; (4) green color in Kucha costume

It is a very interesting example of the costume influence of a small and not powerful at all but highly cultural Kucha Oasis on Turks. What can the reason for such effect be? The influence on some groups of Turks or even some important persons only? Political immigration? Military alliance? Gifts of clothing made for a specific occasion? We do not know.

Fifth, on the Silk Road, Khotan played a big role not only in cotton and also silk textile manufacture (silk production appeared here beginning from 419 A.D. due to a dynasty marriage). It was the only region with mass production of headwear for sale abroad (up to the turn of the 9th–10th cc., when it was also known for the Arab Caliphate). Unfortunately, the exact types of headwear is not identified, still we may suppose they were simple forms for men. We can see in Khotan depictions [Fig. 7 (1)], whereas Kashgar (疏勒) was famous for its braided hats (Si-yu-ki 1906).

Sixth, though the both oases were located within important routes of the Silk Road, the wide usage of imported fabrics and motifs on them is characteristic of more southern Khotan (probably, because in the 1st–3rd cc it was nearer the ways from such mighty states as the Kushan Empire and Parthian Iran and close to them in the main languages). Particularly, in early Sampula necropolis trousers and skirts were made from fabrics of western production or had western plots including people in foreign clothes [Fig. 7 (2)] .

Khotan in the 1st–3rd cc was next to the vast Iranian World. I believe, just because of that, the remains of ancient Saka-Scythian "animal style" were preserved here in clothes décor for a very long time (the same tendency may be seen for much more western nomadic Sarmatians and Alans). The color scale of Khotan early textile was rather diverse reminding northern neighbours of Pazyryk Culture [Fig. 7 (3)].

I have considered the possible influence of Kucha on Khotan and western oases of Xinjiang so far. But sometimes we can suppose a reverse impact of the Khotan costume on Kucha. A bright example of that is the element popular in the Iranian World (Scythians of Ukraine, Sarmatians of Hungary, Parthians of

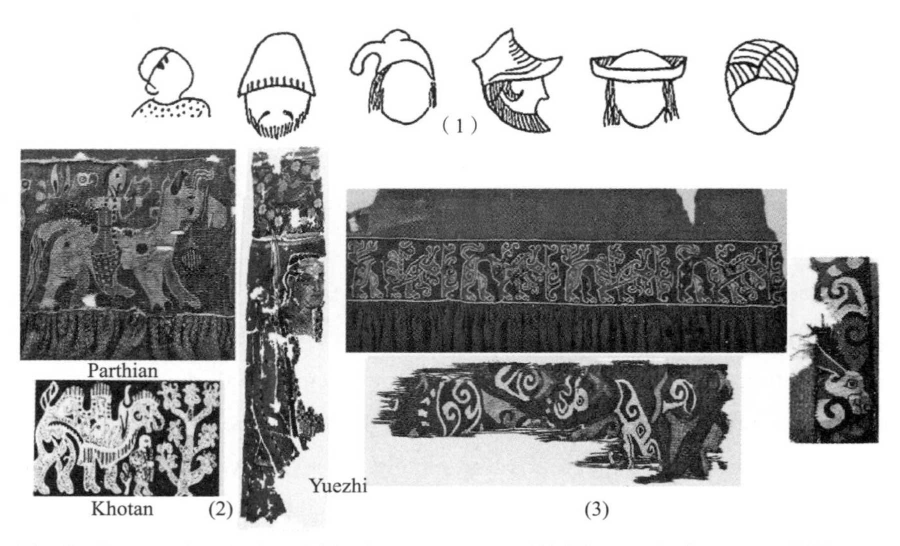

Fig. 7 Some elements of Khotan costume: (1) The typical types of Khotan headwear, probably, used for export; (2) the usage of imported fabrics and motifs on them in Khotan (Sampula); (3) Textile design of "Scythian animal style" of the previous Saka-Scythian epoch in Khotan (Sampula)

South Iran, Yuezhi or Kushans of Bactria) and known in it earlier than others (since the 4th cc. BCE up to the 3rd–4th cc. CE), garments with 6 small triangular projections at equal intervals along the skirt's edge. Later on, both in Kucha (Kizil) and Khotan (Subashi), the same as in more western Ustrushana, male and female short upper garments had already twelve of such projections [Fig. 8 (1)]. Here the question arises as to why the number of such sharp projections is always a multiple of three: 6 or 12? I have answered this question with the help of ancient beliefs of the most northern people on the planet—the Nganasans in the Far North, Siberia. Shamans of this ethnos (with population of 800 people only) preserved an interesting ancient tradition. While shaman's relatives were making a costume for him (on the whole, the costume symbolized a sacred animal—a polar deer). The main part of it was obligatory cut of two deer skins. The deer should be not a domestic animal but a wild one (Prokofieva 1971) [Fig. 8 (2)]. At the same time, Nganasans, according to the tradition, sewed three analogous triangular projections on the skirt of rectangular parts of the garment. The projections imitated the remains of skin removed from the legs and tail. Besides, the spine of the deer to be was imitated ornamentally on the back of the future costume. Thus, a series of triangle projections, 6 or 12

in number, attached to the bottom of the hem once imitated ancient tailoring design of sacred animal's skin. It is evident that such clothes have extended a ritual character. Later, these elements of design spread as far as Western Europe as the décor.

The costume of Khotan is known to us due to clothes from necropolises and series of images from the earlier period of the 1st–4th cc. CE, and this fact

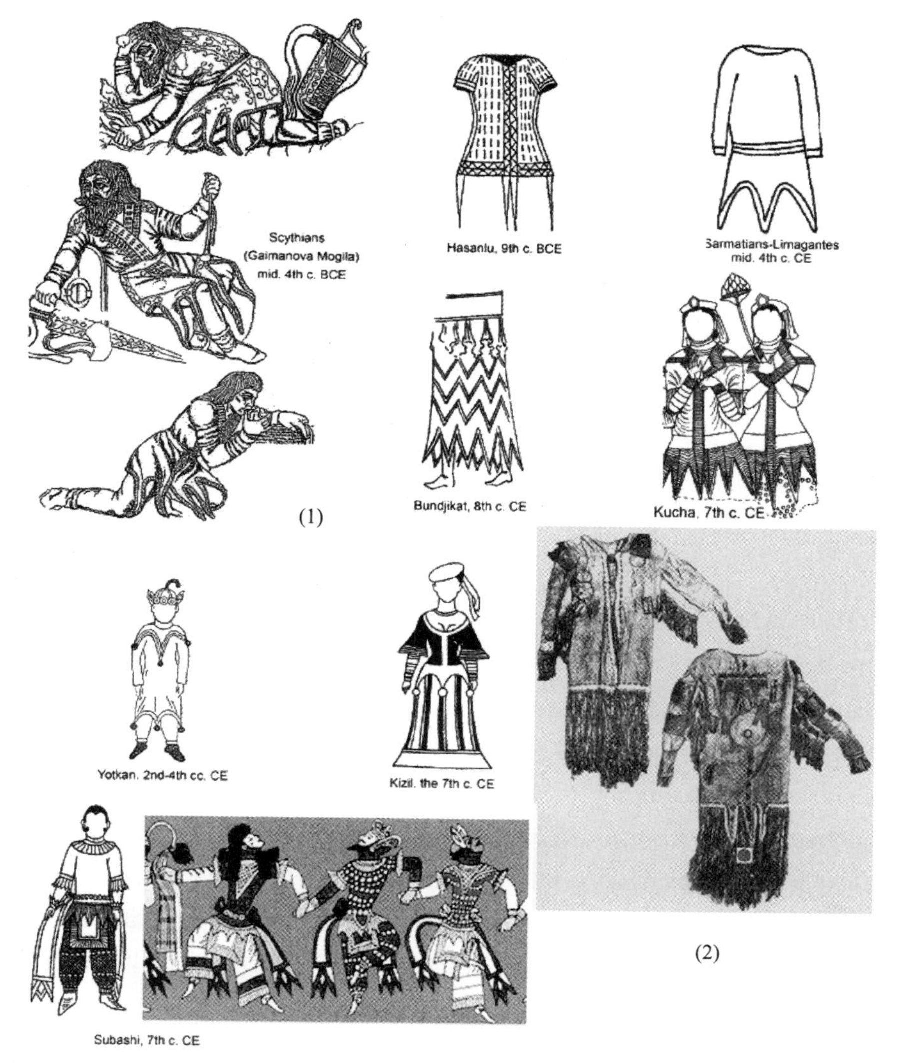

Fig. 8 The garments with 6 or 12 small triangular projections at equal intervals along the skirt' s edge: (1) the ancient and early medieval Iranian and Tokharian peoples; (2) shamanic costume of Nganasan peoples (after Prokofieva 1971)

allows us to find out its evolution process. Comparing these two periods, we note that colors and female coiffure changed dramatically. Upright collars lost their popularity. Sleeved coats now have long sleeves for both genders (not only for women). The waist was accentuated. Vertical stripes on skirts replaced horizontal ones.

In the 6th–8th cc, by the beginning of Middle Ages, Kucha was at the western boundary line of the zone where clothes without shoulder seams were spread. Clothes with detachable hem were characteristic for Kucha, if compared to Khotan. In this Oasis, seams attaching sleeves to the bodice were usually accentuated. On the whole, there was a typical for Central Asia general tendency to substitute upper clothes worn over the head by garments cut from top to bottom and fastened. In the female costume, such substitution took place earlier than in the male one. Such garments were not wrapped but buttoned excluding the upper (breast) part.

The manner to decorate the edges of short sleeves of upper shirts with pleated stripes is known for nobility [Fig. 9 (1)]. Placing of lapels at the collars of male upper garments reflected their social status. Most important aristocrats usually had two lapels [Fig. 9 (2)].

Noble warriors of lower rank, court dancers and musicians had one lapel at the right side; and they also wore belts with a line of metallic plates and a dagger [Fig. 10 (1)]. Two types of shoulder pelerines were worn in Kucha. One type was of local origin. Female pelerines were of three colors, and male ones were monochromic only [Fig. 10 (2)]. Special shoulder medallions differed in form for men (round or drop-shaped) and women (heart-shaped). The beret with two bands was a common headwear for both men and women. A semispherical female headwear is rather original. Its top is decorated with an imitation of a tower with a band through it [(Fig. 10 (3)]. The color scale of the Kucha costume is quite peculiar: the combination of three 'cold colrs'—so-called pastel colors (green, grayish-blue and grey with rare white and black) is predominant.

Fig. 9 Kucha costume details: (1) nobles' manner to decorate the edges of short sleeves of upper shirts with pleated stripes; (2) the aristocratic garments with two lapels

Fig. 10 Kucha costume details: (1) noble warriors of lower rank, court dancers and musicians with one lapel at the right side; (2) shoulder pelerines of local type. Female pelerines were of three colors, male ones were monochromic only; (3) a semispherical female headwear

First, there is an abundance of detailed female depictions (earlier colored images of Turks depict, excluding coins from Chach or Tashkent Oasis, exclusively 'male's world'). Before getting married, girls gathered hair in a topknot and covered it with red silk [Fig. 11 (1)]. Of all female headwear, 'sumuje' worn by noble women and, according to 'Songshi' data, used in ritual performances, presents a special interest [Fig. 11 (2)]. It is a small flat hat, varnished red and joined with two small braids which are tied with a bow at the top. The rest of the hair at that are gathered into two large knots at the sides of the head and covered with translucent black fabric with embroidery fenghuang (凤凰) and clouds. One of Uygur Nestorian Christian women in the church at the western gate to Qocho has an interesting type of headwear. It is a complexly and originally tied turban [Fig. 11 (3)]. An analogous headwear was discovered on a Turkic male statue from Altan-Ŝaandal in Mongolia, and we have no ground to identify its origin in the Turkic culture as connected with Arabic Islamic influence or early Indian tradition.

Second, the following fact presents interest: up to the beginning of the 20th century Uygur women preserved two types of headwear documented in materials of the Pre-Mongolian Uygur state in Turfan: a ball-shaped one flattened from the poles (type 6 after Lyudmila Chvyr') in Kirish paintings, and also small hemispherical hats made of four sectors (type 1 after Chvyr'; later on, all their surface was embroidered with gold thread). There exists an authentic artifact from Qocho ruins [Fig. 11 (4)]. The hats preserved as peculiarities of local groups.

Third, monochromic (red) clothes domineered or the fabric had red basic background (excluding the early stage of Manichaeism/ 摩尼教) [Fig. 12 (1)]; however, in some situations noble people put on garments made of polychromic fabric (evidently, made of local one of high quality). The 12th cc became the most important boundary line in the history of Uygur costume. It was the time when, according to Chinese sources and depictions (for example, the Uygur prince in cave 409), black color started to prevail [Fig. 12 (2)]. Clothes made of dark colored fabric (black, grey, brown, dark-blue)were preserved by Uygurs up to the early 20th cc.

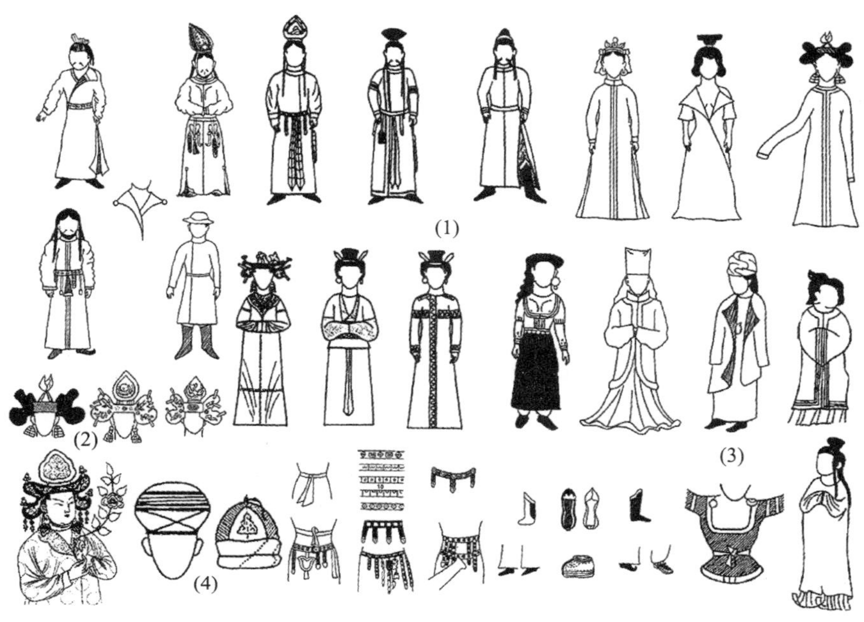

Fig. 11 (1) Turkic Turfan Uygurs' costume elements in the late 9th—11th cc.; (2) female 'sumuje' headdress; (3) Turban one of Uygur Nestorian Christian women, the church at the western gate to Qocho; (4) Two types of headwear documented in materials of the Pre-Mongolian Uygur women preserved up to the 20th cc: ① a ball-shaped headdress flattened from the poles (Kirish paintings); ② a small hemispherical hat made of four sectors.

Fig. 12 Color of Uygur costume of the 9th—12th cc.:(1) The monochromic (red) clothes in the early Uiguria or the fabric had red basic background; (2) Uygur ruler in black costume, Dunhuang, the 12th cc

Fourth, there may be observed a big variety of hair style and headwear for both genders (Fig. 13). It reflected not only social hierarchy but also a complex clan and tribal composition of the Pre-Mongolian Uygur Kingdom. Our studies have shown that the headdress (which was often fastened with a red lace under the chin) was the most important visual symbol of belonging to one of the steps in social hierarchy for men from all six groups of nobility. For the highest group (following the ruler — idiqut himself) — the members of qaghan' clan and other best aristocratic familes, including ministers, a high sharp-topped, complex in its form and basically red or yellow headdress of complete form was a characteristic costume element [Fig. 14, (1)]. The belt, braided of green laces was decorated with round gold plaques (there were two small bags and two pen-cases attached to it). The coat with red hue and with very long sleeves (usually fastened at the wrists with a small button), footwear — black or white high boots, hairstyle — 4–6 long braids and a small beard often parted in the middle.

Less important personages were usually depicted as managers at banquettes or participants of falconry. Their headdresses (fastened with a lace under the chin) have a hemispherical form [Fig. 14 (2)]. They worn either two braids or their hair was cut in a direct line above the forehead. Their long sleeved coats were also made of polychromic fabric. They evidently wore two belts at a time: the upper one was made from red fabric of Chinese type, tied in a bow at the body's front; and the lower one made of leather, decorated with plates and having 4–6 additional hanging straps. Two groups of men of a lower rank serve the needs of the ruler and the people of the highest rank (serve as court musicians, hold fans and parasols above them, carry their ceremonial poles). Their headdresses are a trident-shaped one on a high cylinder base and a small hat with a lense-shaped top of black or white. In the first case their belts also have 406 additional hanging straps (with diamond-shaped plaques). In the second case, they wear one, two or four braids and simple long-sleeved coats [Fig. 14 (3)].

Fifth, analogous to German peoples in Western and Northern Europe, in Uygur Qocho male hairstyle of long braids — from 2 to 12 — turned out to be a

Fig. 13 The types of Uygur hair style and headwear for both genders

Fig. 14 Headress types: (1) the high sharp-topped, complex in its form and basically red or yellow headdress of the members of qaghan (idikut) clan and other best aristocratic familes; including ministers, used a high sharp-topped, complex in its form and basically red or yellow headdress; (2) the headdress less important personages (the managers at banquettes or participants of falconry); (3) Two types of headdresses of the court musicians, hold fans and parasols for the nobles above them, carry their ceremonial poles; the top of black or white color

privilege of nobility, whereas common men started to wear short haircut and shave beards. Commoners' clothing was shorter (knee-length) and belted with a common rope or a strap.

Sixth, the Uygur costume complex of the 9th–11th cc underwent a certain Chinazation. It manifested itself in usage of Chinese fabric, wrapping of garments from the left to the right, wearing long sleeved coats of Chinese type with considerably widening sleeves, wearing Chinese textile belts with long hanging widening ends by both genders and female belts tightening breasts, usage of the specific Chinese male headdress pu'tou（幞头）(by both noble men and commoners), wearing some types of Chinese coiffeur, also shoes and stockings of Chinese type by women, and a rather wide usage of Chinese ornamental motifs.

Bibliography

[1] D. Keller, R. Schorta. Fabulous Creatures from the Desert Sands. Central Asian woolen Textiles from the Second Century BC to the Second Century AD[M]. Riggisberg: ABEGG-Stiftung, 2001.

[2] Gropp G. Archäeologische Funde aus Khotan, Chinesich-Ostturkestan[M]. Bremen: F. Rover, 1974.

[3] Grünwedel A. Altbuddhistische Kultstäatten in Chinesich Turkistan[M]. Berlin: D. Reimer, 1912.

[4] Le Coq A. Bilderatlas zur Kunst und Kulturgeschichte Mittelasiaens[M]. Berlin: D. Reimer, 1925.

[5] Si-yu-ki. Buddhist Records of the Western World (transl. by S. Beal). Vols. 1-2. London：Coronet Books Inc.,2004.

[6] Prokofieva E.D. Shamanic Costume of the Peoples of Siberia (Shamanskie kostiumy narodov Sibiri), Sbornik Muzeia Antropologii i Ethnographii (Collection of the Museum of Anthropology and Ethnology) [M]. Leningrad, 1971: 5-100.

[7] Yatsenko S.A. Costume [Kostium] (Chapter 3), in Eastern Turkestan in the Antiquity and the Early Middle Ages. Architecture. Art. Costume (Vostocnyi Turkestan v devnosti i rannem srednevekovie. Arkhitektura. Iskusstvo. Kostyum). Vol. 4. (Ed. by B.A. Litvinsky) [M]. Moscow: Vostochnaia literatura, 2000.

[8] Yatsenko S.A. Early Turks: Male Costume in the Chinese Art. Second half of the 6th–first half of the 8th Century. (Images of 'Others'), Transoxiana [OL]. Buenos Aires:2009.

个人简介

Inoue Masaru

中亚丝绸之路壁画的著名研究专家

井上豪先生不仅擅长素描绘画，也是中亚丝绸之路壁画的著名研究专家，早年从日本早稻田大学博士毕业，现为日本秋田公立美术大学教授，参与出版了《东洋美术史论丛书》《丝绸之路辞典》《东大寺美术史研究》《禅与人类文明研究》《亚洲佛教美术论集》等著作。井上豪先生对东方壁画的研究十分深入、执着，特别是在中国西域壁画研究领域具有一定的学术影响力。自20世纪90年代起，井上豪先生就对中国西域石窟进行了广泛的考察与研究，截至2016年，共发表关于中国西域石窟的论文、调研、考察报告50余篇，并翻译、编辑了《新疆东部地区出土早期器物的初步分析》《丝绸之路石窟壁画保存》《关于在敦煌窟壁画修复的日中共同研究报告》等。

井上豪先生关于东方壁画的研究教学主要从图像学角度出发，以图像（图谱）分析结合调研为主要形式，并涵盖壁画复制、修复、东方壁画教学方法论等内容，非常独特，具有借鉴作用。

Inoue Masaru, skilled in drawing and painting, is a renowned researcher in the Silk Road murals of Central Asian. Graduated as a Ph.D. student from Waseda University, he now works as a professor at Akita University of Arts. He is a co-publisher of several books including *Japanese Art History, Dictionary of the Silk Road, and Todaiji Temple Art History* etc. Dr. Inoue Masaru has profound and consistent interest in the research of Oriental murals; he holds academic influence in the research circle of caves in western regions of China. Since 1990s, Dr. Inoue Masaru has made extensive investigation and research in the caves of west regions in China, about which he has published around 50 works; he also has related translation works including *Preliminary Analysis on Early Stage Unearthed Utensils in Eastern Region of Xinjiang and Preservation of Murals Along the Silk Road*.

His main research methodology of the Oriental murals starts with images(illustrations) analysis, combining with the on-the-spot research, and covers other teaching and research approaches of murals copy and restore, unique with reference value.

龟兹石窟壁画的供养人和 6 世纪中亚及中国的服饰

Donor Figure of Kucha Murals and Costume of Central Asian and Chinese in 6th Century

龟兹石窟壁画的供养图，是为了纪念出资修建石窟的信徒而绘制的。供养人图和其他佛教绘画不一样，是表现当时现实的龟兹风情的艺术作品。通过仔细观察这些壁画，可以了解当时丝绸之路上的服装文化。

第 8 窟的供养人，是龟兹供养人图最有代表性的例子（图 1）。图中的人都穿着长袍。

图 1　克孜尔第 8 窟供养人

这是第 189 窟的供养人（图 2），供养人两侧的领子都翻开。第 8 窟的供养人只翻开领子的一侧。

这是第 184 窟的供养人（图 3），上面有两个人领子翻开，下面四个人领子束紧。这种服装在西域是常见的衣服，在中亚的壁画中有许多例子。

图 2　克孜尔第 189 窟供养人

图 3　克孜尔第 184 窟供养人

这是巴米扬西大佛龛供养人（图 4），他两侧的领子也是翻开的。

塔吉克斯坦·片治肯特住宅遗址壁画（图 5）中，有三个人的领子也翻开。古代伊朗的传统衣服也是长袍。

2 世纪的纪帕提亚国王像（图 6）穿着圆领的长袍。8 世纪来到中国的中亚人也穿这种衣服。不仅是胡人，汉人也穿这种衣服，说明这是当时国际上的流行服饰。

西域长袍中最经典的是克孜尔第 224 窟涅槃图里的哀悼者所穿的

图 4　巴米扬西大佛龛供养人

图5　塔吉克斯坦·片治肯特住宅遗址壁画

图6　哈特拉出土纪帕提亚国王像（2世纪）

长袍。壁画上描绘了世界民众一起哀悼的场面。这里用服装样式的不同来表示各个民族的区别。壁画中间的几个人是佛教绘画里常见的仙人（图7），其他的几个人是凡人，穿着当时人们的实用服装。

图8壁画中的服装领子的一侧是翻开的。这件衣服的最大特点是领子和后背带有飘带。这个特点与前面的第8窟供养人一样。他们的领口束紧时，全部都用飘带在脖子上的一侧系紧。

图7

让我们来看一下这个衣服领子的样式，这是法雅兹泰珀寺院遗址壁画（图 9），时期是 4–5 世纪。我们可以知道，这种领口有飘带的服装样式早在 4–5 世纪的吐火罗斯坦地区就有。

图 8

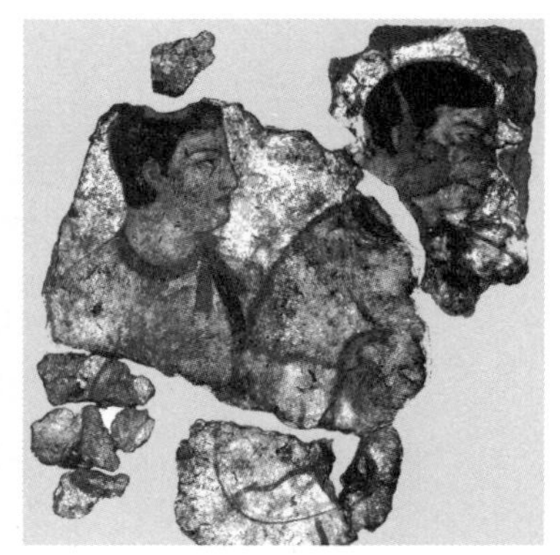
图 9 法雅兹泰珀（Faiyaz-tepe）寺院遗址壁画（4–5 世纪）

图 10 里的人物服装领子一侧翻开，领尖部分是束紧领子用的扣子。

俄罗斯高加索地区出土的贵族衣（图 11）的肩部是用扣子扣紧领口。日本正仓院也收集了很多 8 世纪同样样式的衣服（图 12），也是用扣子扣紧领口。正仓院衣服和西域衣服领口交叉左右相反，领尖各有一个扣子，衣服里的侧面和肩膀一侧各有扣环一个。还有另外一个例子，肩膀的左

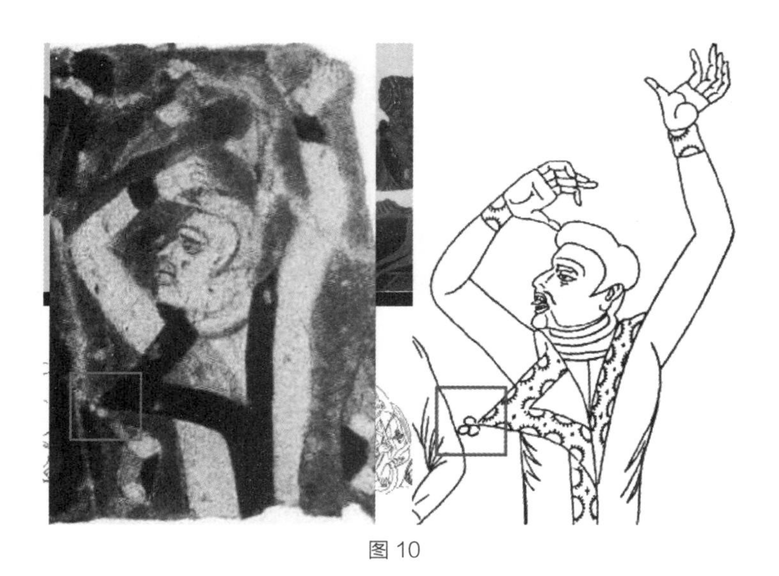
图 10

右、里外侧各有一个扣环，领口束紧。上面的衣服领口只能翻开一侧，下面的衣服领子两侧都可以翻开。这样的领子样式叫作盘领。

图 11

图 12 正仓院衣裳 · 缬缬布袍

我们再看一下这个人物（图 13）。他穿的是两侧盘领，但领间没有扣子和飘带，是不能束紧的样式，这种样式发现了真品。这是俄罗斯高加索古坟墓出土品（图 14），但是领口的左右交叉方式不正确，应该是这种交叉方式，胸部有几个扣子，领子没有扣子。所以说领子敞开不束领长袍确实存在。原来长袍的领子应该束紧，并且是圆领的，但是中亚西域一带流行领子敞开不束紧，所以出现了这种样式的服装。

下面我们看这两个人物（图 15），他们的衣服样式不一样，没有领子。这种样式的衣服叫垂领，也可以叫交领。游牧民族这种衣服的样式

图 13

图 14

图15

比较多。这是北魏的鲜卑族供养人，他们穿的都是垂领长袍。外蒙出土的匈奴服装也是垂领的大长袍。斯基泰战士文壶——公元前4世纪的文物上，人物也着垂领服装。

通过总结（图16）可知，壁画里有四种服装样式：（1）飘带束紧的一侧盘领；（2）扣子束紧的一侧盘领；（3）不能束紧而敞开的两侧盘领；（4）垂领。

飘带束紧的样式与4-5世纪吐火罗斯坦流行的服装相同，不能束紧的敞开样式与8-9世纪在高加索地区流行的服装相同。

领子的束紧方法也在不断地进化，最原始的方法是用飘带束紧，用扣子束紧是一个进化形式，领子敞开是受新的流行影响，从中可以看出西域服装的历史变迁。

领子敞开的流行是如何开始的？请看下面的示例。撒马尔罕阿夫拉西阿卜遗址壁画中描写了粟特王公的婚礼场面。壁画中出现了很多人物。大部分人穿盘领长袍，其中所有粟特贵族的领子都是束紧的，有几个敞开领子的，他们都是突厥人。这个时代的中亚大部分绿洲地区都被突厥人控制，一般的游牧民族特点被当地文化所融合。突厥人的衣服本来是垂领，后来由于受到了粟特人的影响，穿了盘领长袍，因为不适应脖子

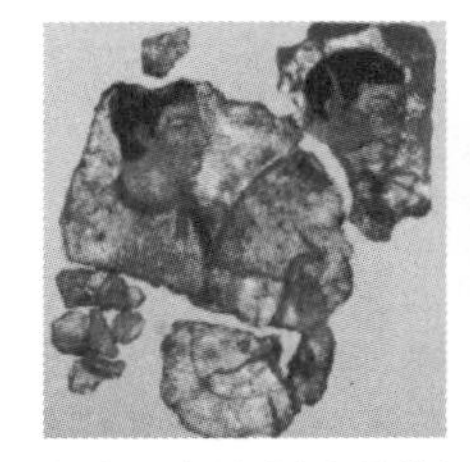
（1）一侧盘领（飘带）

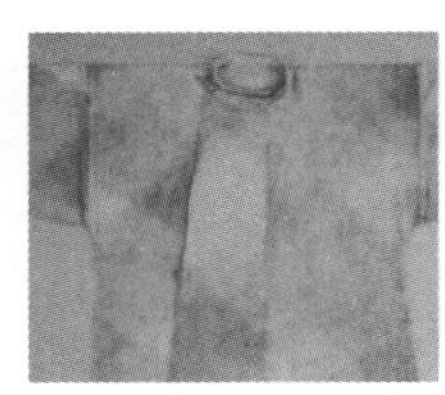
（2）一侧盘领（扣子）

（3）两侧盘领（不能束紧）

（4）垂领

图 16

被束紧而将领子敞开。这种穿法逐渐在突厥人中开始流行，慢慢扩展到西域绿洲，这是我的看法。游牧民族在丝绸之路文化传播上有很大的贡献，中国盘领长袍的流行也是受到了他们的影响。

我们来看中国的敦煌壁画，敦煌壁画上最早出现盘领长袍的例子是西魏的第 285 窟。以前中国美术描绘的游牧民族都穿垂领长袍，叫作胡服。西魏时期开始出现了盘领长袍，从此以后盘领胡服渐渐成为主流。这时期的胡服有两种，285 窟壁画中有几种垂领的胡服（图 17）。盘领长袍从西魏到北周已经完全普及，在汉人中也开始普及使用。

下面比较两个时代的壁画。图 18 是西魏第 285 窟的故事壁画“五百强盗归佛因缘”，官军打败了山里的盗贼，盗贼受罚，被挖去眼珠，流放到山里，在他们失望痛苦时，释迦牟尼给盗贼们宣讲佛法，他们听法之后改邪归正，眼睛恢复正常。战争场面中，盗贼和军官穿的都是胡服，只有官军的指挥官和制裁盗贼的裁判官穿着汉服。但是在盗贼改邪归正的场面中，所有人都穿汉服。由此，我们可以看出汉服与胡服有身份上的差别。着汉服表示有身份、地位高的人，着胡服表示地位低。

然而，这种歧视到了北周就消失了。这是北周 296 窟壁画（图 19），

图17　敦煌第285窟·沙弥守戒自杀因缘

图18　敦煌第285窟北壁五百强盗归佛因缘

也是“五百强盗归佛因缘”。无论是在战争场面还是在改邪归正的场面中，盗贼、指挥官、裁判官都身着胡服，已经没有身份上的差别，汉服人物和胡服人物完全平等。国王在宫殿时穿汉服，出去打猎就穿胡服，对胡服的偏见完全消失，可以看出汉人已经完全认可接受了胡服。这些盘领长袍规定为宫廷衣服，也就是隋唐的常服。

图 19　敦煌第 285 窟北辟五百强盗归佛因缘

北周时期偏见消失是因为受到了突厥影响。552 年，突厥打败柔然，成为草原霸主，和西魏的关系就非常紧密良好，特别是军事关系紧密，联手攻击青海的吐谷浑（图 20）。557 年，西魏王朝演变为北周，北周建国后，与突厥的关系也一直稳定良好。当时的历史背景是西魏和东魏对立，北齐和北周对立，都需要借助突厥的力量抗衡对方。所以北齐也想要和突厥联姻，搅乱北周和突厥的婚约。结果没有成功，最后北周和突厥联合攻击北齐。在这期间，突厥的力量变得越来越强大，地位越来越高，所以汉人对胡服的偏见渐渐消失，汉服与胡服的边界也渐渐消失，胡服融入了汉人的生活中。

可是，敦煌 285 窟是在突厥强大之前建窟，中国最早的盘领长袍不能说是受突厥影响，那么是受哪个民族影响呢？

第 285 窟的壁画描绘的是盗贼被绑起来受罚的场面（图 21），地面上衣服散乱。这些服装的领尖有飘带，领口束紧，这样就和前面看的一样，可以说是 4–5 世纪吐火罗斯坦流行的服装。敦煌 285 窟供养人题记里有几个“滑”姓的家族，“滑”姓是代表厌达人的姓氏，他们是 5–6 世纪控制中亚绿洲的游牧民族，562 年被突厥打败。吐火罗斯坦是厌达人的根据地，所以最早传到中国的盘领长袍是原始的盘领束紧的款式，可能是厌达人从龟兹传到吐火罗斯坦的服装。6 世纪后半期由突厥传到粟特人的衣服，应

突厥与北朝两国的关系

西魏 × 东魏	539	敦煌第 285 窟开凿
	551	西魏的公主降嫁突厥
	552	突厥打败柔然，成为草原的霸主 西魏文帝驾崩，突厥可汗亲自来吊唁
	556	西魏突厥联合军攻击吐谷浑
北周 × 北齐	557	北周建国
	562	突厥萨珊联合军抗击厌达，征服中亚一带
	564	突厥的公主和北周订立婚约，北齐也要联姻，搅乱婚约
	567	北周突厥联合军攻击北齐

↓

需要借助突厥的力量 ——→ 突厥的地位强化

图 20

图 21

该是扣子束紧的长袍。唐朝石雕的突厥人穿的也是扣子束紧的长袍。

我们所关心的中亚和中国服装样式的演变，都可以在龟兹壁画上得到确认，可以说龟兹壁画确实是体现丝绸之路文化交流的重要世界遗产。

个人简介

程应奋

新疆艾德莱斯研发推广中心设计总监，中国十佳时装设计师，新疆十佳服装设计师

毕业于香港理工大学服装与纺织品专业，文学硕士，新疆轻工职业技术学院服装设计师专业带头人，新疆艾德莱斯研发推广中心设计总监。高级服装设计师，中国十佳时装设计师，新疆十佳服装设计师，中国服装设计师协会艺术委员会委员，新疆维吾尔自治区服装（服饰）行业协会副会长，新疆服装设计师协会副会长，新疆维吾尔自治区级服装设计技能大师工作室主持人。多年来一直致力于西域民族服饰文化的研究，举办个人作品专场发布会10余场，四次在中国国际时装周做以艾德莱斯为主题的作品发布。

She graduated from HKPU (The Hong Kong Polytechnic University), M.A. Fashion and Textile. She is anacademic leader of Fashion Design Major at Xinjiang Industry Technical College, and Design Director at Xinjiang Atlas R&D Center. She has got the honor of Top Ten Chinese Fashion Designer, and Top Ten Xinjiang Fashion Designer. She is also a member of Art Committee, China Fashion Association, vice-chairman of Xinjiang Uygur Autonomus Region Fashion Industry Association, and vice-chairman of Xinjiang Fashion Designer Association. She host of Fashion Skill Master Studio of Xinjiang. Her consistent research is in ethnical fashion culture of Xinjiang. She has so far held around ten fashion shows and 4 fashion shows themed Atlas during the International Fashion Week in China.

艾德莱斯——非遗活在当下

Atlas — the Living Intangible Culture

西域很有特色的一种纺织品，名字是艾德莱斯。

美若彩虹，灿若云霞的艾德莱斯丝绸，是新疆和田地区特色传统手工艺制品，它出现距今已有近 3000 年了，被称为古丝绸之路上的活化石。一片花绸的诞生，要经过扎经染色等多道工艺（图 1）。艾德莱斯丝绸使用我国古老的扎经染色法工艺，用草根、核桃皮等食品浸泡的植物颜料染色，其色彩缤纷绚丽，虚实相间，对比强烈，图案简洁抽象又不失生动，风格上唯美浪漫。它再现了大自然中光和色的美感，吸取东西方文化之精髓（图 2）。

在大家的认知中，古丝绸之路上新疆的和田地区在一个很重要的位置，恰恰又是东西方汇集的一个地方，所以这个地方受到了诸多文明的影响：古罗马、古巴比伦、中国的文化、印度的文化都在这个地方留过痕迹，同时又有很多宗教对这个地方产生了影响，比如，佛教、伊斯兰教等。所以这个面料有将近 2000 年的历史，受到了很多文化的影响，带着“国际血统”（图 3）。

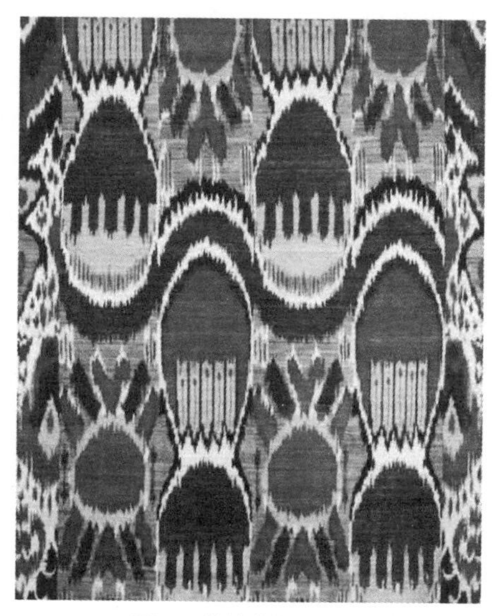

图 1　艾德莱斯丝绸 1

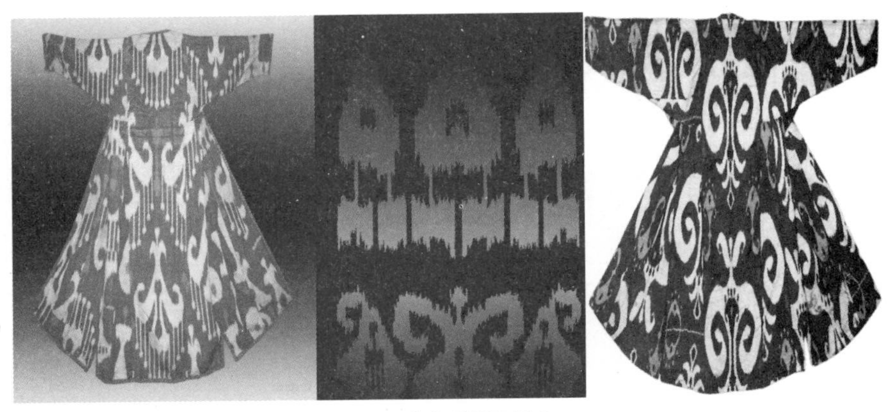

图 2 艾德莱斯丝绸 2

图 3 艾德莱斯丝绸 3

目前这个面料在整个的中亚地区都有使用。研究的过程中，我们发现艾德莱斯丝绸最主要的特点，是扎经染色。学过纺织的同学可能知道，我国经常会用到这个扎经染的方法，与蜡染和扎染是一个派系。

扎经染色是把面料扎完了以后放到染料去染，面料是在经线上做设计，我们所有的图案全部从这个经线的设计开始，经过经线的设计之后去制造。在将近 2000 年的历史中，曾经在印度、伊朗、周边的乌兹别克斯坦、哈萨克斯坦都有大量的织物的生产，并且都有非常鼎盛的时期，相互都有影响。

艾德莱斯丝绸因为是手工制造，所以通常情况下它的宽度是 45cm，

为了人的手能够借助梭子。艺术楼一进门的地方，右手边有艾德莱斯丝绸面料展，嘉宾们还可去楼下再去参观。在这个 45cm 左右的面料上面，设计师进行图案的设计。在整个的制造过程中间，艾德莱斯丝绸都经过了扎经染色整经织绸一道道的工序。在工序的过程中，我们过去都会使用植物染料和矿物染料，但是随着我们时代的发展，有更多的这个化学染料介入之后，很多工匠也开始用化学原料染色，图 4 展示的就是这个缫丝的过程。

图4　缫丝 1

过去和田地区有大量桑树种植，所以丝绸的产量也是非常大。图 5 中的老人是艾德莱斯丝绸非物质文化遗产的传承人之一，她在对丝线进行整理，它的核心技术，就是扎经染色。

它是以每一个家族为支系，因此每一个家族的图案纹样特点会不一样，又因为是口口相传，所以没有文字记载。纹样设计的核心部分由男工匠来做，在经线上去做设计，然后把需要的纹样全部设计好之后，妇女或者是男士都可以去做，把不需要染色的部分扎紧。在过去最早的时候用玉米皮，现在有塑料薄膜，所以他们会用塑料薄膜把主要的部分，

图5　缫丝2

也就是把不染色的部分缠绕起来，缠绕以后把剩下的部分量好之后从木框上取下来晾干，晾干之后第二次再进行染色。纹样有几个颜色，它就需要染几次，如图6~图8所示。

图6　扎经染色1

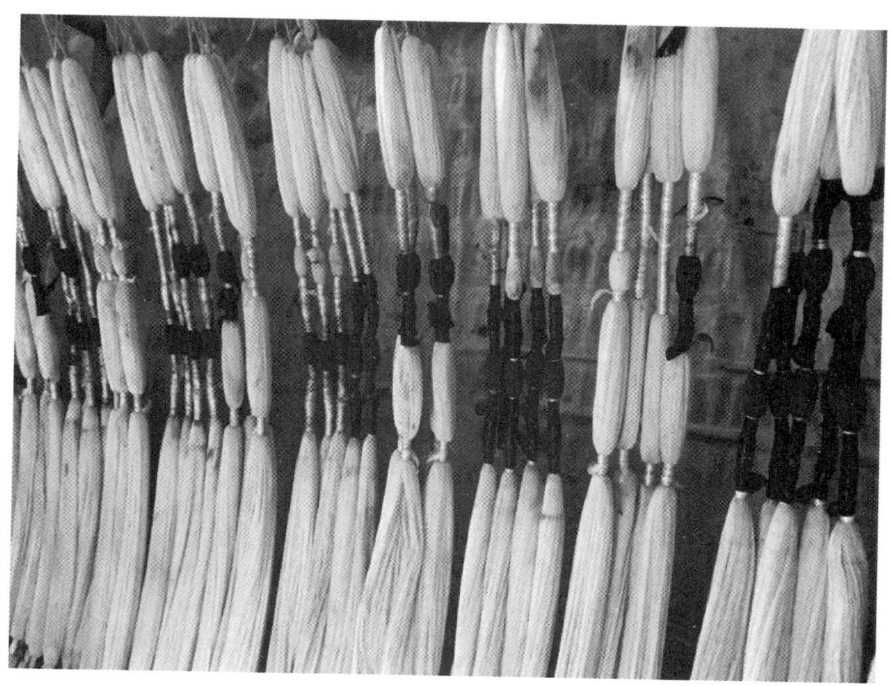

图7　扎经染色2

图8　扎经染色3

在染色的过程中，因为每一个工匠手的松紧程度会不一样，所以颜色的渐变是一个很奇妙的过程，颜色的渐变和印花或是织法、绣法表达出来的感觉不同。

图9　扎经染色4

艾德莱斯图案是经过特殊的扎经染色工艺形成的。但是不是所有的纹样都可以通过扎经染色来表现，那么也正是因为这样，扎经染色形成了独特的图案风格。手工操作使得花纹图样中每一个花形多少都有些不一样，这也造成了后续制作服装时会有困惑——经常在花形的匹配上花很大的工夫。艾德莱斯图案当时也受到伊斯兰教的影响。伊斯兰教徒不喜欢人物或者是人物的眼睛出现在面料上，所以图案基本上就是日常用具和植物比较多，如图10~图13所示。

图10　艾德莱斯丝绸植物图案

图 11　艾德莱斯丝绸饰物图案

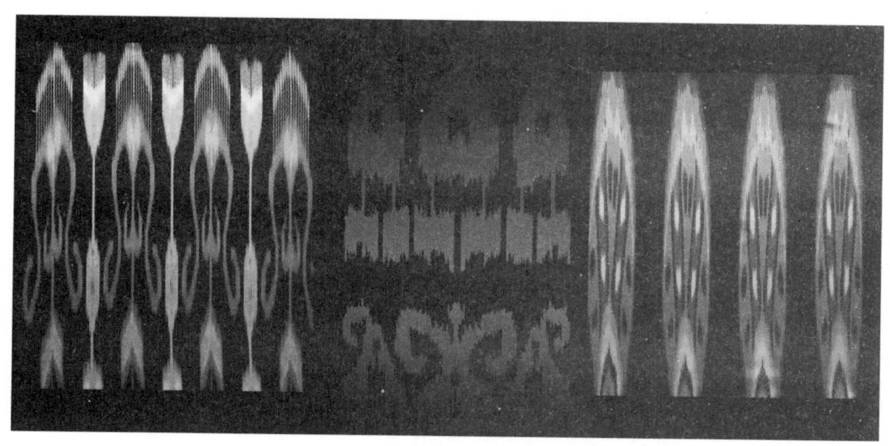

图 12　艾德莱斯丝绸工具图案

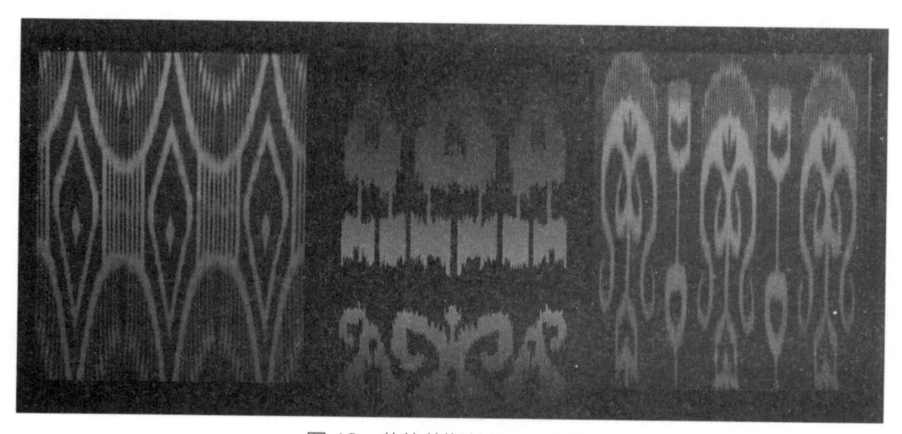

图 13　艾德莱斯丝绸乐器图案

艾德莱斯丝绸在我国新疆地区广为流传，有将近2000年的历史。现在看到的是乌兹别克斯坦的艾德莱斯丝绸，品种更丰富，他们有棉、棉麻、毛的多样品种。

艾德莱斯丝绸出产区在我国新疆和田的吉亚乡，在吉亚乡有1.13万的妇女在从事艾德莱斯丝绸的制造，所以很多人都是靠艾德莱斯丝绸摆脱贫穷的。在我关注艾德莱斯丝绸之前，也就是将近十年前的时候，因为艾德莱斯丝绸的制造是口口相传，每一家便是一个支系。但当艾德莱斯丝绸消费量变低之后，很多人家就不再制作了，孩子们从事其他的工作，一个支系也就消失了。

当我开始关注的时候，我就开始收集各种不同面料，慢慢地也得到了政府的支持。现在这个产业慢慢开始复苏，过去关掉的家庭支系又开始重操旧业了，但是大多数的支系中工匠技术比较好的人的年龄基本都在70岁以上，新生代从事这个行业的人数比较少。所以非遗活在当下的含义，其实更重要的就是让更多的艾德莱斯丝绸走入人们的生活，消费艾德莱斯丝绸，那么更多的家庭支系中会有更多的动力来从事生产。这些年我们受到了来自政府越来越多关注之后，也有更多的设计力量开始加入了艾德莱斯丝绸的拯救之中。

其实大家如果注意到的话就可以发现，国际上这个纹样运用特别广泛。艾德莱斯这个词是从维吾尔语翻译过来的，目前为止中国和周边的国家在纺织品方面有很多的交流，即他们的产品过来，我们的产品带过去，所以"一带一路"的文化是互融互通，没有障碍的，因为我们喜欢的东西他们也喜欢。目前艾德莱斯丝绸在25个国家和地区都有销售，见图14~图16。

最近这两年，艾德莱斯做了很多的宣传和推广活动。也有在政府的引导下举办大型的活动。在亚欧时装周、中国国际时装周、上海时装周当中都有设计师一直在参与关于艾德莱斯的发布，希望有更多的人关注和喜欢艾德莱斯纹样。这个非遗和其他非遗有一点不同的是它可以形成一个产业。例如，某一些非遗定位的只是人群中很少的客户，但是艾德莱斯有实际的生产意义，艾德莱斯可以走进每个人的生活。我们曾经拿这些面料去伊朗做展示，非常受欢迎。在座的一位女士带着的那条围巾

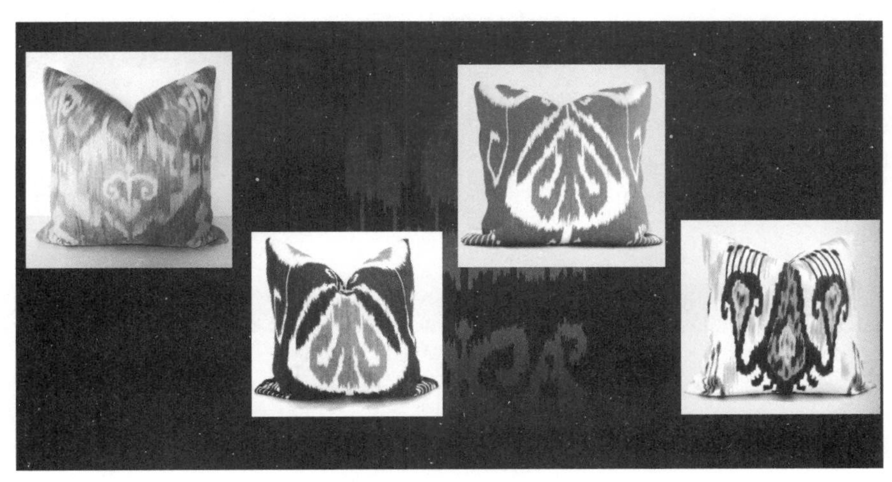

图 14　艾德莱斯丝绸产品 1

图 15　艾德莱斯丝绸产品 2

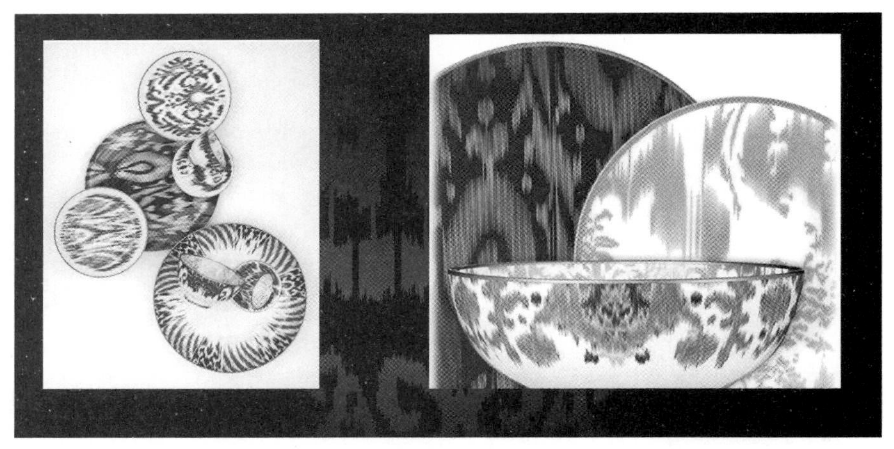

图 16　艾德莱斯丝绸产品 3

就是艾德莱斯。

在深圳文博会的时候，我碰到了一个法国专家，他建议我们一定要深入地研究艾德莱斯，因为它带着国际血统。因为它是有持续生命力的，不是某一阶段有，下一阶段就没有了。随着国际流行趋势的发展，可以把图案做成豹纹、花纹格等，如果这一季的流行变了，我们可以把颜色或者是图案做改变。紧跟流行趋势，她说这样会使艾德莱斯丝绸有持续的生命力。

艾德莱斯丝绸作为 21 世纪最后的丝绸手工制作工艺，经历了千年的发展，已经不仅仅是一种织物，而是一种承载着历史积淀的西域民族文化艺术，这种美丽的带着匠人体温的丝绸面料，深深地吸引着世界各地的人，越来越多的国际设计师用艾德莱斯纹样做设计。这种文化的温度以及艾德莱斯的人文情怀，使得艾德莱斯所承载的时尚在国际舞台上更有力量。

最后我想分享一个小故事，我很早就开始搜集这个面料，但是每一次我去到和田，都会见到不同的艾德莱斯支系和面料。有一次我和我的同伴去了和田一个老爷爷传承人家里，我们看到从没有见过的面料，特别激动，但是当时因为我要赶飞机，我就跟他说："快快快把它帮我打包，我要拿走"。我也要给他付钱，但是老爷爷并不理我，帮我挑好面料

后把它放在地上，然后他就跪在铺在地上的地毯上，而我那时急切的希望他将面料放进我的行李箱，随后乘车赶飞机。但无论我怎么催他，他都不听。他跪在地下，把面料放得平平的，按照他们的方法把一块一块的面料全部排列好，随后就像诵经一般，念念有词。但我听不懂他在说什么，他念完了以后，用双手帮我把面料放在行李箱里，然后我就一下被震撼到了。艾德莱斯丝绸的确是带有人的体温的面料，我们要心怀敬畏地对待它。因为它已流传了 2000 年，它的能量，它的神秘和意义可能无法一个个地被解读，但确是值得大家去喜欢的。

个人简介

贺阳

北京服装学院民族服饰博物馆馆长，博士生导师

北京服装学院教授，民族服饰博物馆馆长，博士生导师，北京市人大代表。曾设计2008北京奥运会、残奥会系列服装，设计航天部“神七”“神九”宇航员舱内工作服、睡袋，“神舟十一号”宇航员地面保暖装置等。设计60周年国庆北京志愿者服装、群众游行国旗、国徽方阵服装，设计第十六届亚运会颁奖礼仪服装。荣获科学技术部科技奥运先进个人，中国创新设计红星奖特别奖等多项荣誉。纺织之光教师奖、北京市师德先进个人、北京市宣传文化系统“四个一批”人才。

Dr. Heyang is a professor at Beijing Institute of Fashion Technology, curator of the Ethnic Costume Museum, supervisor of Ph. D. students, and delegate of Beijing NPC. She designed for Beijing 2008 Olympics and Paralympics, Shenzhou VII Spaceship, Shenzhou IX Spaceship, and Shenzhou XI Spaceship (Astronomer's work clothes, sleeping bag and warming equipment). She also designed for Beijing volunteers costume, parade national flag, and national emblem parade costume at 60th anniversary of the National Day, as well as the Miss Etiquette's costume at 16th Asian Games. She has been awarded as the advanced Olympic personnel by the bureau of Technology, the Distinguished award of Chinese Creative Design (Red Star Award), Teaching Award of the Light of Textile and Beijing Advanced Moral Person etc.

传统苗族服饰结构中的智慧

Wisdom in Traditional Costume Structure of the Miao Ethnic Group

传统苗族服饰结构充满智慧。

这是两层土布（图 1），用针缝起来，把它们缝合在一起，会有一种看起来很硬的感觉，这样就改变了布的质感，让它的剪裁更加的立体，更加有挺括的造型感。

这是一位老人，她就穿着这样剪裁的衣服，后面很好看，有一个褶皱（图 2）。

图 3 是它的效果图。图中的模特穿上这件衣服，再配上这个头饰，就有一种特别的平衡感。这件服装，并没有要求穿衣者的腰很细，而是通过一条大约有五六米的布，把腰部一圈一圈地缠住，缠住之后，腰其实还是很肥，但是看起来很紧致，可以感觉到腰部很细很有力量，裙子

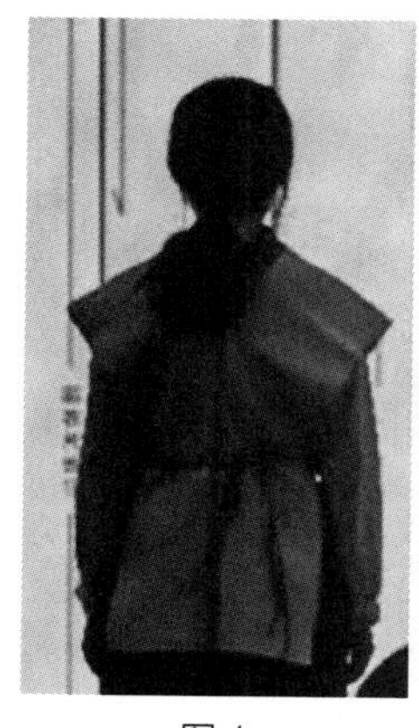

图 1

图 2

图3

很大、很漂亮。这些结都是衣服上的结，她穿了好几件衣服，都是衣服上打好的结。

图4里的衣服，是平常收集、用来上课的。这款服装是方形，肩部的线条前片短，后片长，它通过穿着的方式让这款服装有造型感和立体的效果。

贵州息烽县青山苗族上衣

衣服前后连肩，领子缝合于横开口处，所有裁片都是矩形，领口直条包边。后背有蜡染和刺绣装饰。后背“背牌”挑花，带子到前胸交叉，将前片包裹身体，侧缝无缝合，包裹合体，带子于后面系结，此举相当于立裁手法。宽大的后片先拉平，将两侧向外中对折，系腰带固定。衣服呈现上大下小的造型。通过这两个步骤，完成了衣服从平面到立体的塑造，使衣服具有建筑感

中国传统服装用尽量少的剪裁，体现“节用”和“惜物”，通过穿着方式，完成立体造型，穿着的二次设计，使衣服适应性强，不同高矮胖瘦都合适

装饰体现了“慎术”，不过度装饰，适可而止，少即是多。又一次证明好的设计在民间

衣服结构图

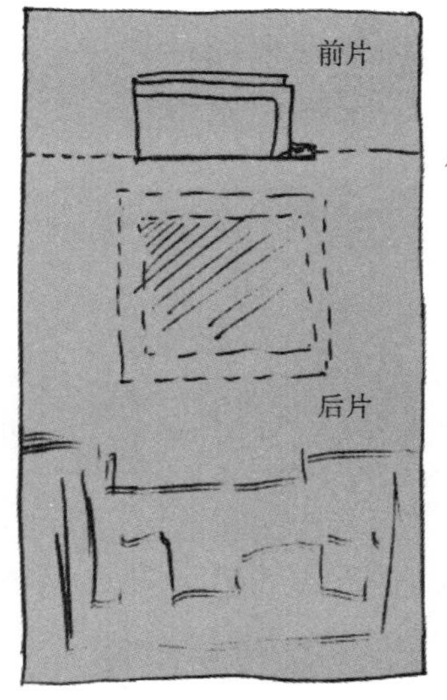

图4

图 5 是这款服装的实物，是当地人的服装。穿的时候把短的这边穿在前面（图 6）。服装上有一个很大的领子，领子有一个白色的边，穿完之后把前面部分，就是前片，用裙子把它束在腰里，这样腰就显得很细，后面要折过来。

其实穿着方式很多，还有从后面像折扇一样把布折上去的，穿着方式可以自由发挥，但其实也就几种，因为只有这几种方式穿出来最好看。所以有很多传统的东西保持很长时间，几百年、几千年，都是同一个款式。它一直在经营图案的摆放，穿着方式的整理，尺寸比例的剪裁。之所以能把一件衣服做得特别漂亮，是因为一直在经营考究比例关系。因

（1）正面

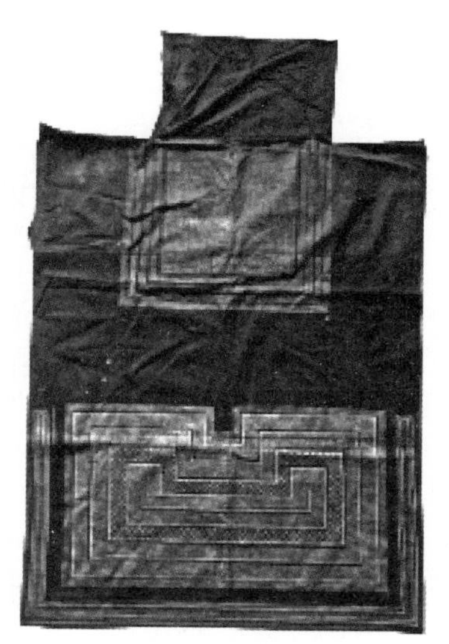

（2）背面

图 5

图 6

为好的设计都是通过不断的使用演化而来的，就像一只很好的碗，又好拿、又不烫手、装的东西又多、又不容易洒，都是由长期的使用和改进得到的。所以，好的设计在民间，这是需要学习的。

穿完之后，还要用一个围腰（图 7），把它绑起来。绑起来之后，有一块长方形的布，布上有两根带子，两根带子在胸前交叉绑起来，穿完就是图 8 的这种效果。这个美丽的袖子其实就只是一块长方形的布做的。

图 7

图 8

图9是从后面看的效果，其实它还有很多好看的腰带垂下来，这其实也是一种非常古老的方式。苗族的衣服几乎没有扣子，都是用系带和缠绕的方式来做的，这与中国传统的服饰风格一脉相承。

图9

图10是另一件衣服，它也是用了长方形的布。这个领子也是长方形，但是，这个长方形在前面有交叉。穿上以后，领子也是长方形的两条布。年轻人可以穿，老年人也可以穿。

以前，苗族人都是将布料染一些简单朴素的颜色，但现在他们也喜欢用化学染色的腈纶毛线来做，因为觉得颜色很漂亮，但笔者还是觉得传统的更好看。如果染色的时候没有弄好，就会过于鲜艳，特别突兀，以前他们采用的是植物染色，那些颜色都比较偏灰调，看起来就非常和谐地融合在一起。因为植物染色的过程中通常有很多的杂质，是这种杂质让颜色有了一种含灰的效果，并不是特别纯的颜色，所以它们配在一起都很自然、柔和。

这些都是他们的衣服，他们的一套衣服要穿很多件。图11中的服装已经逐渐淘汰了，剩下的配件都不全，前面还差一个围腰，因为这个裙

图 10

子前面开衩，要用围腰把它挡住，也不再有头饰了。图 11 中模特肩上是一块方形布。但是现在他们已经放弃了这种刺绣的方法，因为觉得在视觉上不够吸引人，就做了银饰，做了更多的装饰，但我还是觉得原来那种比较好看。

想要了解更多的信息，请关注博物馆的网址 (http://www.biftmuseum.com)。可以为大家提供更多的信息。

图 11

个人简介

张慧琴

北京服装学院语言文化学院院长，硕士生导师

张慧琴，上海外国语大学英语语言与文学博士，北京服装学院语言文化学院院长，教授，中国翻译协会专家会员。研究方向为中外服饰文化差异、跨文化交际和应用语言学。主持国家社会科学基金项目和北京市哲社重点项目等，入选北京市“长城学者”培养计划，多次荣获教学科研成果奖，并在《中国翻译》《中国外语》以及《外语与外语教学》等期刊发表论文40余篇，出版有关服饰文化、跨文化交际，服饰文化与大学英语教学融合的专著、译作10余部。

Zhang Huiqin got her Ph.D. of English language and literature in Shanghai International Studies University; She is the Dean of School of Language and Culture, Beijing Institute of Fashion Technology; Editorial Board Member, Journal of Higher Education in BIFT and Expert Member of Chinese, Translators Association. Her research Interest is fashion, cross-culture communication and applied linguistic. She was in the charging of several research programs funded by National Social Science Fund, Beijing Municipal Education Commission-funded and Beijing municipal government, Participated “the Great Wall scholar” program in Beijing, and was awarded for teaching and research achievements. She came out Journal Articles more than 40, including “On Confucianism reflected in the description of clothing in the Analects of Confucius”, “On translation study of Hebao in A Dream in Red Mansions from the perspective of cultural translation”, “An inquiry into the strategies employed in the translation of the dressing etiquettes” etc. She published more than 10 works on cross- culture fashion communication, harmonious translation strategies and combining fashion culture into college English teaching.

《论语》中以"礼"服人的哲学思辨

Confucius' Thoughts Reflected in the Function of Clothes in *the Analects of Confucius*

Now I will start with my paper. The topic is *Confucius' Thoughts Reflected in the Functions of Clothes in the Analects of Confucius*.

As our research, most studies on the Analects of Confucius focus on his ideas on education thoughts, political ideas, and moral ethics. Few have paid attention to his comments on clothing. Today, I would like to study the color, design, pattern and fabric about clothing to reflect the etiquette of his time, the social stratification, social obligation, and codes of conducts, and further to argue that these clothing codes and styles should be part of Confucian thinking.

The first part is about the Analects of Confucius and clothing. Analects of Confucius stipulates a series of codes of conduct for the people to follow in their daily life. These codes are consistent with Confucius' political ideas and social ethics. The codes make sure the people could live peacefully and stay in their social position. The codes of clothing, part of general codes of conduct, focuses more on maintaining the hierarchical aspect of the society and individual behaviors. Based on the core spirit of Analects of Confucius, the clothing plays an important role in building up the harmonious society at that time.

First, ceremony and etiquette concerning attire may keep society as a whole in perfect order. It requires that everyone should act according to their social obligations. The social class was judged differently in terms of color, fabric, style and pattern. Second, ceremony and customs dictate the social norms, which means that clothes must conform to the requirements of etiquette even under different situations. The proper wardrobe at the proper time could even transform and define the man's reputation well. Third, attire functions as etiquette in social situations, helping people cooperate with each other harmoniously.

Now the second part. Confucianism reflected in the description of clothing in James Legge's version of Analects of Confucius. This section includes how does clothing reflect Confucius' idea of hierarchical thinking of society, his idealism of social harmony, and the harmony between human and natural worlds while examining and analyzing specific examples from the Analects of Confucius translated by James Legge? First, let's look at the colors in clothes. Colors are important to traditional Chinese philosophical thought according to those who believe in the Taoist philosophy of 'Wu Xing'. The colors blue, red, yellow, white and black are associated with wood, fire, earth, metal and water. To Confucius, the regulation of color in clothing was a part of ideas of harmony between human and natural world. There are some examples to illustrate it much more clearly. The first example is the color of the collar. According to *the Analects of Confucius*, "The superior man did not use a deep purple, or a puce color, in the ornaments of his dress. Even in his undress, he did not wear anything of a red or reddish color." After the era of slavery, in the federal society, color of clothes did not only serve with practical purpose, but also had the function of attracting others' attention. Colors also had to distinguish the social class between the noble and lower ranks, and as a symbol of class, dark red, yellow green, black and white were seen as noble colors. And for the main-stream colors, between the green and the white, the dark and the red, all these intermediate colors were considered as low rank colors. According to the customs of the western, the bright red and bright purple was the noble colors for noble use. In Confucius' opinion, the collar of decoration of gentlemen's clothes should not be in dark green or was supposed to use black on the inside of the clothes as a decoration of sleeves; besides, he could not use red or purple cloth for regular household purpose (Fig.1).

Fig. 1

The third part is the etiquette in clothing. In this example, you will find that, “When a person was ill and the prince came to visit him, he had his head to the east, made his court robes be spread over him, and drew his girdle across them.” The example indicates that even a junior official was ill, and stayed in bed in the period of ancient China, he must wear the formal attire, including his court dress and the belt, in case of the emperor’s visit, so as to show his proper respect to the emperor. In other words, his attire must match his position in front of the emperor (Fig.2).

The fourth part is clothing style. In this part, it is mentioned that, “On the first day of every month, everyone put on ‘Weishang’, his court robes, and presented himself at court to meet the emperor.” “Weishang” is also for occasions of ceremonies. The style of “Weishang” resulted in using much more cloth than the normal clothes, but it was necessary for the sake of etiquette. At the announcement of the new moon, he put on his court robes and left for court in full court regalia (Fig.3).

Fig. 2

Fig. 3

The fifth part. I'm going to say something about the clothing fabric. It is said in the Analects that, "When fasting, he thought it necessary to have his clothes brightly clean and made of linen cloth." When preparing themselves for the sacrificing ceremony, they must wear some good clothes of linen, which means that one must dress properly in different situations, and darker colors were much more favored than lighter ones in traditional Chinese clothing. So the main color in the ceremony was dark rather than light. Lighter colored clothes were worn more frequently by people for their everyday use (Fig.4).

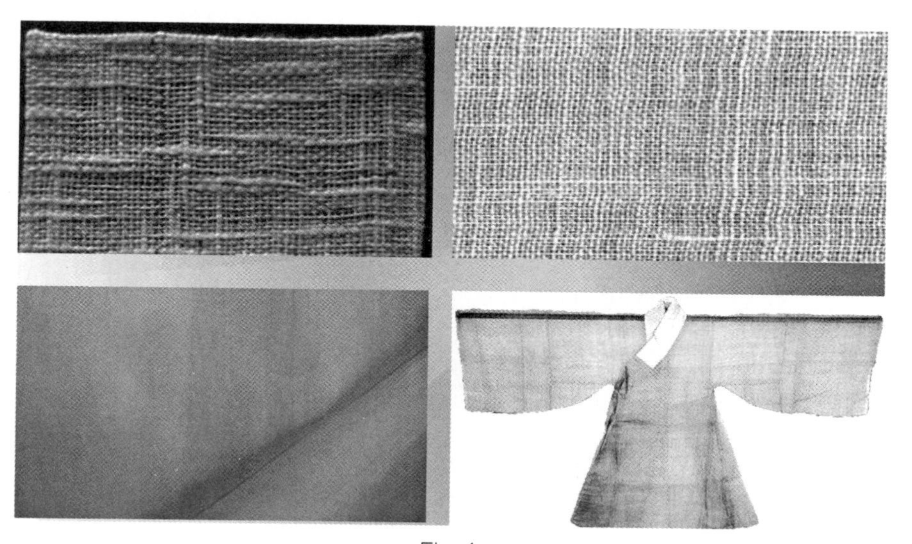

Fig. 4

The sixth part is clothing ornaments. "When staying at home, he used thick furs of the fox or the badger without paying attention to personal ornaments. When put off mourning, he wore all the appendages of the girdle." The decorations not only added the attraction of personal beauty, but also signified their social status. People often chose jade and jewelry in the shape of knives and swords as the personal ornaments. But the most common material, which was typically jade. However, when the mourning was over, the appendages of the girdle, which were lucky talismans (Fig.5).

The seventh part is the rule of clothing. The master said, "Where the solid qualities are in excess of accomplishments, we have rusticity; where the

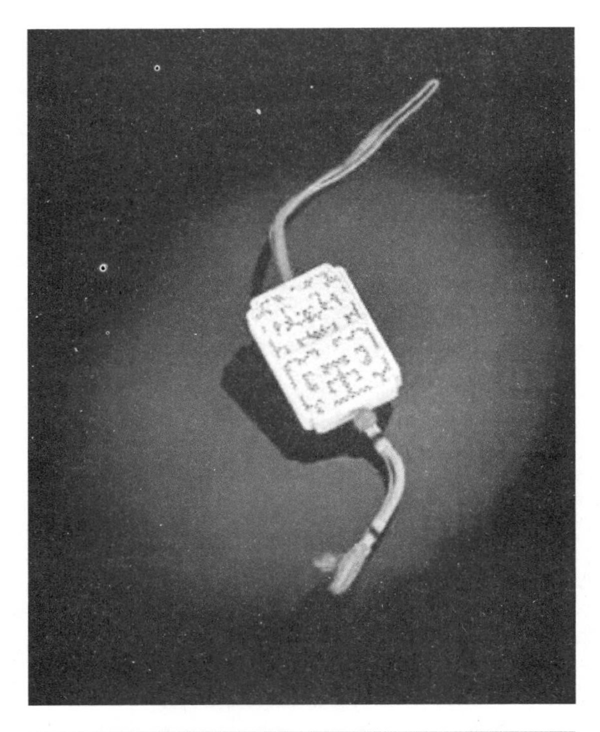

Fig. 5

accomplishments are in excess of the solid qualities, we have the manners of a clerk; When the accomplishments and solid qualities are equally blended, we then have the man of virtue." Confucius not only advocated the beauty of clothing form, but also formulated the relationship between form and content in cultivating the attire of a true gentleman.

In conclusion, the *Analects of Confucius* is not only a book of his political ideas, but also an early book that mentions the dressing codes and etiquettes in ancient China. Confucius' sayings on clothing both reflect and reinforce his hierarchical sense of society and idealism of social harmony. The clothing codes of the later Chinese dynasties, though different from one another, all continued these dual functions. That's all. Thank you!

个人简介

李艳

首都师范大学北京语言产业研究中心执行主任，博士，教授

李艳，首都师范大学北京语言产业研究中心执行主任，博士，教授，主持国家社科项目、北京市重点项目，多次享受国家留学基金委资助，先后赴美国、英国、中国台湾等大学访问学习。研究方向为语言产业、语言消费、文化传播。

本文发表在《语言文字应用》2017 年第 4 期，为 2016 年度国家社科基金项目“‘一带一路’建设中的‘语言消费’新问题及其对策研究”（项目号 16BYY053）的阶段成果。

Li Yan, the executive director of the Beijing language industry research center of Capital Normal University, Ph. D., Professor, presided over the national social science project, the key project of Beijing City, and has been supported by the National Research Committee of the State Fund for many times and has visited universities in the United States, Britain and Taiwan, China. The research orientation is language industry, language consumption and cultural dissemination.

This article was published in the fourth issue of *Language Application* in 2017. It is a part of achievements of the 2016 National Social Science Fund Project “New Problems in Language Consumption in the Belt and Road Construction” and its countermeasures (Item 16BYY053).

语言消费：基本理论问题与亟待搭建的研究框架

Language Consumption: Basic Theoretical Issues and Research Framework

目前，国内外对“语言消费”问题都还尚未有系统的研究，究其原因：一是随着语言生活与经济行为的互动发展，学界对语言产品（服务）的界定、特性的认识处于不断完善之中，相应影响到“语言消费”问题的研究进程；二是“语言消费”行为具有较强的渗透性，难以切分、剥离，这给测量与分析带来了困难；三是“语言消费”作为其他消费的基础，其消费动机、行为较之物质产品及其他文化产品的消费更为复杂，西方经济学中的消费理论、西方社会学家关于文化消费的一些经典论述并不能简单地用于“语言消费”问题的研究中。在多种因素的作用下，国外语言经济学界还尚未将“语言消费”作为研究的重要组成部分；国内对“语言消费”的研究也处于起步阶段。

尽管研究的难度客观存在，但是，“语言消费”不仅关系着整个社会的语言生活质量，也直接影响着文化的传承与传播，并直接或间接对经济发展产生影响，因此，“语言消费”研究的意义与迫切性不言而喻。

一、何为“语言消费”——“语言消费”内涵与外延近十年间的宽窄之变

研究者对“语言消费”的界定在近十年间呈现出明显的变化，这一变化与国内“语言产业”研究的开展以及学界对语言产品（服务）特性

的认识不断完善有关。国内"语言产业"研究系统的开展始于2010年，这一年也相应成为"语言消费"研究发展的标志性时间点。

（一）随着对"语言产品（服务）"界定的清晰，"语言消费"的外延也相应从窄到宽

2010年之前，研究者将"语言产品"界定为语言类教科书和音像制品；将"语言服务"界定为语言培训、语言翻译服务。并在此基础上，将围绕"语言学习"展开的消费活动称为"语言消费"，包括购买语言学习资料、参加语言培训和测试等。也就是说，在这一时间段，受对"语言产品"和"语言服务"界定较为狭窄的影响，研究者将"语言消费"的范围仅划定为"语言学习"方面的消费。这一界定在今天看来具有明显的局限性，但客观来看，或许也是与当时业界与学界对"语言产品（服务）"的认知相吻合的。

2010年之后，随着国内学者对"语言产业"研究的关注，"语言产品"的概念也逐渐清晰，以语言为核心要素或主导要素，以满足某种语言需求为目标的产品形态，包括语言出版、语言培训、语言翻译、语言测试、语言文字信息处理、语言艺术、语言康复、语言会展等业态的产品，都被归入"语言产品（服务）"，对这些语言产品（服务）的消费，都属于"语言消费"。亦即，对以上"语言产业"所属各个业态所提供的所有语言产品（服务）的消费，都属于语言消费。

如果说这是从"语言产业"的视角对"语言消费"内涵与外延的界定的话，那么，李宇明教授对"语言服务"的界定，则又进一步拓展了"语言消费"的边界。他认为，"利用语言（包括文字）、语言知识、语言艺术、语言技术、语言标准、语言数据、语言产品等语言的所有衍生品，来满足政府、社会及家庭、个人的需求"的服务，都属于"语言服务"。根据这一界定，在语言产业所提供的商业化语言产品（服务）的基础上，由政府、高校、科研院所以及语言事业机构提供的非商业性或者具有"语言福利"性质的语言产品（服务）也都可以作为"语言消费"的对象，如语言标准、语言数据等；而且，除了个体的消费者外，政府部门也可以作为"语言消费"的主体，如对来自科研机构的语言决策咨询方面研究的消费等。

（二）根据研究视域的不同，“语言服务”的主体从语言行业拓展到非语言行业，在一定范围内拓宽了“语言消费”的边界

不论是在语言产业还是语言事业的研究视域下，“语言消费”的对象都须是以语言为核心要素或主导要素，以满足某种语言需求为目标的语言产品（服务）。但是如果将“语言消费”研究拓展到传播学、社会学、经济学等人文社会科学领域，那么，“语言服务”的主体随之相应扩大，可以从语言行业拓展到非语言行业，因为，语言存在于一切交流之中，在任何消费活动中都离不开对语言的使用，包括产品说明等书面语言、服务人员的口头语言等。这些伴随于其他消费活动的语言行为，虽然不同于以语言为核心要素或主导要素的语言产品（服务），但是其对于文化的传播、经济活动的进行、社会的有序发展等都有着直接或间接的影响。

因而，有研究者将金融、交通、医院、饭店、商场以及工商、税务、公安等窗口行业的从业者为“消费者”提供服务时所伴随的语言作为“语言服务”的内容，继而将“消费者”在这些场所购买产品或服务、办理业务时所接受的语言作为“语言消费”的内容。但需要明确的是，消费者在就餐、购物等过程中并不是以满足某种语言需求为目标，而是以餐饮、购物为消费目的，尽管从业人员的“语言服务”会影响消费者的消费体验，但并非其最主要的消费内容。所以，有必要将这种伴随式的“语言消费”与前一种以消费语言产品或服务为目的的“典型性语言消费”进行区分，并且明确其特定的含义与适用的研究范围，避免将“语言消费”概念泛化。

综上，“语言消费”包括了以语言产业为供给主体的“典型性语言消费”和以窗口服务行业为供给主体的“伴随式语言消费”。除此之外，还涵盖了对以政府、非营利性质的科研院所、社会公益机构为供给主体的语言政策、语言文字规范标准、语言教育、语言数据、语言康复等服务的消费，这一类消费可以归入“典型性语言消费”。

二、为何研究"语言消费"——"语言消费"的功能

国内外"语言消费"相关研究的整体数量较少，且近年来对于"语言消费"内涵与外延的界定处于变动之中，故而截至目前尚未见到对语言消费功能的系统论述。综合语言消费主体的消费需求与客体的产品特性，可以将语言消费的功能分为直接功能与延伸功能两个方面。"语言消费"的直接功能包括提升语言能力、感受语言魅力、优化消费体验等三个主要部分，并且，这三个部分又各自具有一定的延伸功能。

（一）直接功能

1. 提升语言能力

"提升语言能力"是"语言消费"最为重要的功能，也是"语言消费"的首要动因，其主要来自于对语言教育培训、语言翻译、语言出版、语言测试、语言康复等产品与服务的消费。"提升语言能力"既包括提升母语能力，也包括提升外语能力。如果根据消费主体的状况和需求进行细分的话，对于语言康复服务的消费旨在使需求者获得、恢复正常的语言交流能力；而以拥有正常语言交流能力者为消费主体的语言培训、语言翻译、语言出版、语言测试产品（服务）的功能在于使需求者拥有更好的语言素质、更高的语言技能以及通过成功的语言转换实现良好的跨语言、跨文化交流。如果以"基本的语言交流能力"为坐标原点的话，那么，位于原点以左的"语言康复服务"和位于原点以右的"其他语言产品（服务）"则分别负载着实现"由无到有""由有到优"的语言能力提升目标。

2. 感受语言魅力

"感受语言魅力"来自于对相声、朗诵、书法、文字创意设计等语言艺术及语言创意产品（服务）的消费。一部分个体、群体在语言消费基本需求的基础上，产生了更高的消费需求，如在广告、命名、饰品及旅游产品设计等行业中对语言文字元素创造性运用的需求，我们可以称之为语言创意需求。例如，对"命名服务"的消费，旨在为机构、组织、产品或者个人取一个意韵深远、朗朗上口的名字，这类语言消费是以语言

文字所具有的发音、字形、意境的美感为前提，通过自我感受、展示分享因语言的独特魅力而令人称道的名字，提升自身的可识别度、形成富有个性的品牌，在此基础上，获得社会评价、经济效益方面的回报；此外，服饰及旅游产品设计中将汉语古文字和少数民族语言文字作为核心创意元素也属于一种语言艺术产品，对这类产品的消费，具有审美功能，并且，其中具有较高收藏价值的产品还将给购买者带来一定的经济回报。

3. 优化消费体验

"优化消费体验"主要是针对来自窗口服务行业的"伴随式语言消费"而言，以城市公交为例，清晰的标识、贴心的提示、得体的答复等有效的语言服务，可以使乘客获得良好的搭乘体验，推及到就餐、就医等其他场景，亦是如此。

语言消费活动具有一定复杂性，在某一个语言消费行为、语言消费过程中，可能同时涉及对多种语言产品（服务）的消费；也可能某一种语言产品（服务）兼具了不同的功能。因此，对某一语言消费行为及其功能的分析需要结合具体情况来进行判断。以语言文字信息技术产品和语言翻译产品（服务）为例，前者多作为"嵌入式"产品，存在于其他语言产品与服务之中，如输入法、电子词典、语音翻译、语音导航以及医院、银行的智能对话系统等，因而也兼具了多种功能；后者既可以单独存在，也可以融合于语言教育培训、语言出版、语言艺术、语言技术等其他产品或服务之中，因而，在功能上也不尽相同。

（二）延伸功能

"语言消费"的延伸功能包括经济功能、文化功能和社会功能。

1. 经济功能

第一，包括母语和外语在内的语言能力的提升有助于提高人力资本，"语言消费者"个体的语言技能与工资收入有较强的正相关性，对于一个国家而言，亦是如此，"包括母语水平和使用外语的人数、熟练程度"在内的国民总体的语言能力，"是该国人力资本的一个重要组成部分"；同时，语言能力的提升能够增强不同语言群体之间的经贸往来，降低经济活动协调、管理、信息交流的成本。

第二，"典型性语言消费"拉动语言产业发展，可以为国民经济创造可观产值，例如，"瑞士语言的多样性，为瑞士每年创造 500 亿瑞郎的收入，约占瑞士国内生产总值的 10%"，"世界上每年有 1800 万母语非西班牙语的人学习西班牙语，与该语言学习相关联的产业产值每年可达 1500 亿欧元"。根据估算，我国语言产业中的语言翻译、语言培训、语言出版、语言技术、语言测试、语言康复六个行业 2016 年产值合计约为 4190 亿元人民币，在国民生产总值中的占比为 0.56%（此为作者按照一定的计算方法所做的估算）。其中，语言翻译行业所占比值最大，其次为语言培训行业。HCR 慧辰资讯 TMT 互联网研究部针对一、二线城市用户的调查结果显示：超过 65% 的一、二线城市用户过去一年内参加过外语培训，用户从学龄前儿童到不同年龄阶段的成人，语种上以英语为主（超过 80% 的用户参加的是英语语言培训，其次为日语和韩语），线上学习方式逐渐被越来越多的用户所采用。根据艾瑞咨询研究院对 2016 年前三季度国内移动教育企业细分领域融资分布情况的分析，"语言学习"类企业占比为 17.9%，仅次于 K12 教育（23.2%）和职业技能教育（21.1%）。2016 年前三季度国内移动教育企业融资过亿的 13 个案例中，纯"语言学习"类的占了四席。少儿在线外教类企业近一年完成 10 笔融资，获得资金在 24 亿元人民币以上。由北商研究院、北京大学孕婴童产业课题组、互联网教育研究院、未来工场联合发布的《2016 在线教育趋势报告》预测 2017 年在线语言培训市场可达 355 亿元人民币。

第三，窗口服务行业从业者良好的语言服务能力，可以优化顾客的消费体验、增强顾客购买商品的意愿，带动商品的销售，同时，包含语言服务在内的良好的服务能力，也会优化当地的人文形象与环境，有助于促进当地旅游业的发展。并且，研究也表明有相当比例的消费者愿意为"伴随式语言消费"付费，其中，"77.1% 的消费者能够接受占总服务价格 5% 及以下的语言服务价格，40.0% 的消费者能够接受占总服务价格 6%~10% 的语言服务价格"。

2. 文化功能

语言消费与文化消费、语言传播与文化传播是一个互相促动、循环往复的过程。

第一，语言消费有助于拓展文化传播的受众范围。文化的传播必然是借助某一种语言来抵达其受众的，受众的语言能力影响着文化传播的范围与效果，而语言消费正是受众语言能力得以提升的重要途径，包括对母语和其他民族、国家语言的学习和使用。

第二，文化传播有助于增强受众的语言消费意愿。对一个国家的文化或者某一种文化产品感兴趣而开始学习这个国家的语言，在外语学习者中是一个较为常见的现象。如中国功夫电影、日本动画片、韩国电视剧等，是一些青少年学习汉语、日语、韩语的动因之一。不仅这一由跨文化传播而引发的外语学习属于语言消费，实际上，在跨文化消费中也包含着直接的语言消费行为，例如，对影视剧片名、字幕等语言翻译产品的消费等。

第三，受众语言消费意愿的增强，又是以文化传播的拓展与深化为结果的。可以说，不论受众是以文化消费为起点，还是以语言消费为起点，在消费过程中，必然都会经过消费包含语言产品的文化产品（如译制片）或是本身就属于文化产品的语言产品（如外语教材等语言出版物），继而产生对与该文化相关的语言或是该语言相关的文化的兴趣及消费意愿，从而形成新的消费循环，如此往复，最终达到文化传播的目的。

因此，语言消费所具有的文化功能是不言而喻的。简而言之，在“典型性语言消费”中，语言艺术产品的消费本身就属于文化消费，语言技术产品、语言翻译服务等为文化产品的生产与传播提供着技术与语言支持，语言教育培训能够潜移默化地影响学习者对于这一语言所属文化的整体认知；在“伴随式语言消费”中，窗口行业的语言服务水平对于一城、一地，乃至一国文化的建构与传播也有着不容忽视的作用。

3. 社会功能

一些必需的“语言消费”关乎国计民生、关乎社会发展，以语言康复产品与服务的消费为例，据残疾人口普查数据显示，目前，我国言语听觉障碍患者数量超过 3500 万，每位患者的家庭按 3 口人计算，涉及人数超过 1 亿人，因此，语言康复服务能否及时跟上、满足需求，不仅是医学问题、语言学问题，更是一个社会问题。

同时，政府作为语言服务的提供者之一，通过普及国家通用语言文字、保护各民族语言文字、完善语言文字规范标准、规范和推广国家通用手语及盲文、推进语言康复治疗技术开发利用等项工作来着力提升国民语言能力、构建和谐语言生活，而这也正是和谐社会建设的基础。

三、如何研究"语言消费"——亟待搭建的"语言消费"研究框架

西方的消费理论、行为消费理论、消费社会学研究以及国内外关于文化消费、语言经济学的研究可以为"语言消费"研究提供一定的理论基础和方法借鉴。在此基础上，还需要把握语言消费相对于其他消费的差异性、独特性，从而搭建适用于语言消费研究的基本框架。

（一）来自经济学、社会学领域的"消费"理论——"语言消费"研究的理论基础

1. 西方经济学者提出的消费理论

20世纪30年代，凯恩斯的"绝对收入假说理论"认为消费水平取决于绝对收入水平，距此理论问世10余年后，杜森贝里提出了"相对收入假说"，认为消费者的消费支出除了受自身目前收入的影响之外，还受到其他人消费支出的影响以及自身以往"高峰时期"收入的影响，这说明了消费行为具有一定的效仿性，消费习惯一旦形成就会产生一定的惯性。这两种"假说"的共同点都是关注现期收入对当前消费的影响，不考虑未来的预期收入，同时，未涉及不确定性问题。

此后，莫迪利安尼等人提出的"生命周期假说"和弗里德曼的"持久收入假说"在消费函数计算中加入了对"预期收入"的分析。其中，"生命周期假说"认为个人现期消费取决于个人现期收入、预期收入、开始时的资产和个人年龄大小，一个家庭的消费支出与家庭中每个人在其生命周期内消费的理想分布有关；"持久收入假说"认为消费者的消费支出是由其可以预计到的未来收入决定的，并且，消费者即便在现期收入走低的情况下，也可能会采用预支未来收入的方式来保持过去"高峰收入"时期的消费水平与消费习惯。这两种"假说"相对于前两种"假说"

的进步意义在于纳入对“预期收入”因素的分析，但不足在于仍旧未考虑到不确定性因素对消费的影响，由于未来的收入具有不确定性，因此，有必要将该因素加入到对消费行为的分析中。

针对以上“假说”中存在的缺陷，之后的“预防性储蓄假说”涉及未来不确定性因素的分析，认为不确定性越高，人们就会更多地选择储蓄而非消费，来降低可能发生的风险。

对西方经济学者提出的消费函数理论需要客观看待，一是消费函数并不应是一成不变的，时代变迁、经济发展、制度差异等因素都会对人们的消费心理与行为产生影响，相应需要消费函数随之调整；二是西方消费理论对于中国消费者的消费行为的解释力是有限的，我国许多学者的观点认为，西方的传统消费理论无法完全说明中国消费者的行为特征。

2. 西方社会学者对消费问题的研究

西方社会学领域对于消费问题的集中研究较早可见于 20 世纪 60~70 年代，并在 20 世纪 80 年代末、90 年代初得到快速发展，使消费社会学逐渐成为社会学研究的一个分支。其特点在于把消费作为一种社会现象，而不是一种单纯的经济现象，关注影响消费的社会因素，从社会阶层、社会结构的角度去分析消费的动机与行为。

韦伯认为消费与生活方式相关，具有不同生活方式的人有着不同的消费模式与习惯，并且消费方式与生活方式是判断特定阶层地位的重要标志；鲍德里亚认为人们正是通过消费特定的类型的物品来对自己的身份进行界定的；齐美尔认为时尚消费与不同社会群体之间客观存在的风格差异有关，社会中的上层人士不断以新的时尚来与其他阶层进行区隔，而其他阶层的人士则不断通过时尚消费来模仿上层人士的生活与风格；兼具制度经济学家和社会学家身份的凡勃伦是系统研究消费的社会结构意义的第一人，他认为社会成员的生活水准取决于他所隶属的那个阶级所公认的消费水准，为了显示和维持自身的社会地位，进行炫耀性的消费是必需的，否则就有可能被本阶层的其他成员轻视或排斥；布尔迪厄在韦伯的基础上，进一步阐释了消费对于阶层分化的作用，对文化资本的意义与作用进行了强调，认为文化知识的等级体系与消费者的社会等级体系是一致的，在资本、场域、惯习的理论框架中，分析了消费的社会区分功能以及不同阶层存在消费差异的原因。

消费社会学研究的积极意义在于弥补了经济学领域的主流消费理论的研究局限，后者在消费研究中所运用的“代表性消费者”假设忽略了不同社会阶层的消费行为差异，这种差异是客观存在的，不能简单将微观变量加总为宏观变量，还需要加入对社会收入分配、社会阶层地位等因素的分析，因此，将社会阶层地位因素纳入消费函数对于经济学领域的消费研究具有借鉴意义。不过也应看到，消费社会学研究还有一些尚待解决的问题，如为描述社会阶层地位建立可以量化的指标，确定不同阶层消费行为的微观基础等。

3. 国内外关于文化消费的研究

国外学者关于文化消费的研究同样也主要来自于社会学、经济学两大阵营。社会学界对文化消费的研究从法兰克福学派将文化消费视为一种社会控制的手段，到布尔迪厄将文化消费与宏观社会结构的特征、社会阶层的区分关联起来，再至丹尼尔·米勒将文化消费作文化创制的一种过程，呈现出一个从否定到肯定的发展脉络。经济学界对于文化消费的研究多是采用微观经济学的视角和实证研究的方法，在理论运用上，从新古典消费理论到新消费理论，再至提出“理性致瘾”理论和“消费中学习”模型。

其中，运用新古典消费理论研究文化消费的问题在于将文化产品等同于普通消费品，仅运用收入、价格、偏好这三要素来解释文化消费行为，且由于经济学在分析偏好这一要素的形成与影响时，不能像对收入、价格的分析那样驾轻就熟，只好将消费者的偏好假定是固定不变的。在新古典消费理论之后，新消费理论的进步意义在于把文化产品（服务）的特性纳入到效用函数中，认为在收入、价格因素相同的情况下，影响消费者选择的主要因素是其对文化产品特性的感知。但是，新消费理论在对消费者偏好的分析上并未能超越新古典消费理论。

此后，文化消费“理性致瘾”理论认为消费者在以往文化消费体验的基础上形成一定的文化消费品位，并不断积累成为消费者的文化资本，文化资本可以通过外化为一定的文化消费能力来对文化消费意愿与行为产生影响。“消费中学习”模型认为消费者的偏好、品位和文化资本是在

不断地学习中形成的，因此，在对消费者文化消费行为的考察中，需要采用动态研究与过程分析的方式。“理性致瘾”理论和“消费中学习”模型对当前的文化消费和语言消费研究都有一定的借鉴意义。

国内学者对文化消费的研究集中开始于20世纪80年代中后期，在近30年间，研究涉及的内容包括文化消费的含义、特点、功能、分类以及影响因素等。但是，从文献梳理来看，尽管研究者对文化消费的定义日趋达成共识，但对于文化消费所涵盖的范围、所包含的内容却仍存有较大差异。

4. 语言经济学领域的相关研究

根据舒尔茨1962年出版的《教育经济价值》一书，语言学习是教育投资的构成部分之一，也是一种人力资本投资。20世纪70~80年代，语言经济研究者开始关注语言的人力资本属性，认为人们主动获取语言技能的投资有助于经济优势的形成。语言技能可以为人们带来很大的利益回报，语言作为交流工具，可以在消费与生产活动中将个人的经济福利最大化。有些研究发现，双语者或多语者能够更容易地学习其他语言和技能，甚至于一些雇主将是否学习过另一种语言作为评价求职者学习新事物能力的一个标准。国内研究者根据贝克尔·奇斯威克的最优教育模型分析语言（技能）投资的过程，得出这样的结论：从人力资本理论角度来看，“纯粹的经济激励”是人们学习另一门语言的主要动机。

语言经济学研究认为，除了人力资本因素外，语言歧视也会对人们的语言学习投资产生影响，如詹姆斯·莱文森通过对南非的动态调查，发现尽管英语的收入回报率在经济全球化的背景下总体呈上升趋势，但是，不同种族之间的差异较为明显，说明语言的回报会受到种族影响。如果说人力资本和语言歧视是影响语言学习投资的直接因素，那么，语言政策可以看作间接因素。并且，语言政策与语言歧视问题具有一定的因果联系；语言政策与语言歧视最终是通过劳动收入（人力资本）这个杠杆对语言学习的投资发挥作用的。

从梳理来看，虽然语言经济学者的研究没有直接关注或提及语言消费，但是他们所探讨的语言学习、语言技能投资问题，实际上就是对语言教育培训产品（服务）的消费，这里的“投资”可以理解为在语言消费上所投入的资金成本。因此，语言经济学研究中的人力资本理论以及一些模型、方法可以为语言消费研究提供参考。

（二）“语言消费”研究的要素与方法

在“语言消费”研究中，需要根据语言产品（服务）的特性进行分类，在统一分析的基础上，探讨不同类型语言产品（服务）的消费动因、需求、方式以及所适用的研究方法。

国家统计局“文化及相关产业分类”中对文化产品（服务）的界定为我们判断文化消费所涵盖的范围、所包含的内容提供了依据，同时，也有助于我们划定语言消费与文化消费的交叉部分以及语言消费中相对独立、不属于文化消费的部分：一是不包含在文化消费中的语言消费部分，语言教育培训产品（服务）是语言消费的重要内容，并且围绕语言教育培训的开展，也涉及对语言翻译、语言出版、语言测试等产品（服务）的消费。语言教育培训产品（服务）的提供主体为各类教育机构。但是，在国家统计局2012年对文化及相关产业的分类中，删除了“国民教育”部分，因此，与教育相关的这部分语言产品（服务）是语言消费中相对独立存在的部分。二是语言消费与文化消费的交叉部分，语言消费的对象有一部分是包含在文化消费的范围之中的，如“语言出版”隶属于《文化及相关产业的分类（2012）》中的“新闻出版发行服务部分”，“语言艺术”隶属于“文化艺术服务”部分。三是某些种类的语言产品（服务）中，部分属于文化产业的统计范畴，如“语言文字信息处理”产品与服务中，有一部分属于“文化软件”服务和电子快译通、电子记事本、电子词典等“文化用品的生产”。四是某些种类的语言产品（服务）是直接为文化产品生产提供服务的，如“语言翻译”对于“影视节目的制作与发行”。

对于语言消费中相对独立、不属于文化消费的部分，需要根据产品（服务）的特性来确定适当的研究方法。以语言教育培训产品（服务）为例，根据国内社会学者的研究，消费分层受职业分层的影响较小，但与受教育程度、家庭人均收入、家庭类型等相关性较高，特别是其中的受教育程度这一指标，不仅对消费分层具有显著的恒定影响，并且日益成

为普遍公认的合理社会分层体系的参照标准。因此，在“语言消费”研究中，可以结合语言经济学“人力资本”与社会学“消费分层”的研究视角，分析语言教育培训消费对个人收入、人力资本的影响，继而产生的对其他消费以及消费分层的影响；作为一个循环，消费分层又在一定程度上对包括语言教育在内的语言消费需求、行为产生影响。在语言教育培训消费行为的具体分析中，可以借鉴行为经济学、行为消费理论的一些观点和研究方法，行为经济学相比主流经济学来说，更重视社会心理因素的影响，如认为带有估测偏见的人在预测其未来偏好时，会给予当前偏好过大的权重，夸大自己未来的偏好与当前偏好的相似性；行为消费理论所建构的心理消费模型有助于分析影响甚至决定消费决策的社会心理动机，其主要采用的是内省和心理实验的方法。

对于语言消费中与文化消费有交叉，或是隶属于文化消费的部分，可以借鉴以往文化消费研究中的要素设定及研究发现。所涉及的要素包括收入、年龄、性别、家庭成员结构、教育水平、职业身份、替代品价格以及消费所投入的时间成本等。这些变量对文化消费都有着显著的影响，但在具体的消费行为中又呈现出一定的差异性和复杂性：收入对文化消费的影响根据所消费产品（服务）的类型不同会呈现出一定的差异；教育水平和职业身份对文化消费有显著的积极影响，特别是对于表演艺术消费，教育水平是最重要的影响因素；此外，文化消费也呈现出较为显著的地区差异。也有研究通过实证调查，对以上变量在人们文化消费中所起作用按由强到弱进行了排序：收入 (X=0.16)、职业（X=0.124）、婚姻状况（X=0.12）、年龄（X=0.11）、文化程度（X=0.106）、性别（X=0.09）。

综上，对于语言消费中与文化消费有交叉的部分，可以借鉴文化消费研究的分析方法；对于语言消费中相对独立、不属于文化消费的部分，需要对产品（服务）特性以及消费者的消费动因、需要等进行分析，确定其适用的分析方法。同时，借鉴西方经济学、社会学中对影响语言消费、文化消费的宏观环境、收入、社会阶层、文化资本、偏好、消费惯性等要素的研究及其具体的测量方法，结合当前语言产品（服务）消费的特性，确定适当的研究方法。

（三）"语言消费"研究的基本思路与框架

语言消费可以根据消费主体的需求划分为不同的层次，对于个体消费者来说，可以分为以下三个层次：基本语言消费，对应的是消费主体获得基本语言能力的需求；中端语言消费，对应的是获得具有一定竞争力的语言技能和相匹配的人力资本；高端语言消费，对应的是差异化消费需求，旨在强调阶层属性的小众群体。

语言消费研究也可以根据研究纵深度的开掘，划分为三个层次：第一层次的研究内容包括语言消费主体、消费对象、消费需求、消费方式、供给主体、供需状况、供给对策等；第二层次的研究内容包括语言消费需求的形成机制、影响语言消费行为的内部与外部因素、语言消费习惯的稳定程度及其动态变化过程、"理性致瘾"和"消费中学习"在语言消费中是否存在以及是如何作用于语言消费行为的；第三层次的研究内容包括语言消费的现有总体规模及潜在规模、语言消费对于语言产业发展和国民经济发展的推动作用、国家相关语言规划与语言政策，主要计算来自"典型性语言消费"所带来的直接经济效益，在此基础上，对"伴随式语言消费"带来的间接经济效益进行推算（表 1）。

表 1　语言消费研究的基本框架

语言消费研究的三个层次	
第一层	语言消费主体、消费对象、消费需求、消费方式、供给主体、供需状况、供给对策等
第二层	语言消费需求的形成机制、影响语言消费行为的内部与外部因素、语言消费习惯的稳定程度及其动态变化过程、"理性致瘾"和"消费中学习"在语言消费中是否存在及是如何作用于语言消费行为的
第三层	语言消费的现有总体规模及潜在规模、语言消费对于语言产业发展和国民经济发展的推动作用、国家相关语言规划与语言政策

在语言消费研究的三个层次中，第一层是最为基础的研究，是第二层和第三层研究得以进行的前提。当前阶段亟待开展的是第一层的研究，为了便于表述，这里以"一带一路"建设"五通"中的语言消费研究为例，对语言消费第一层的研究进行梳理（表 2）。

表2 “一带一路”建设“五通”中所涉及的语言消费第一层研究的主要内容

“五通”	语言消费主体	语言消费需求	对应的语言产品与服务	供给主体	供给对策
政策沟通	国家、地区各级政府机构及非官方组织	以满足沟通所产生的“语言转换”需求为主	语言翻译	翻译机构、企业、个人	专门的外语人才培养 对外汉语教育
设施联通	道路、交通、通信、能源企业的相关设计、施工、监管、运营方	专业技术方面的语言翻译需求；外派员工与当地员工日常语言交际需求	语言翻译 语言培训	翻译机构、企业、个人；高校、孔子学院（课堂）、语言培训企业	专门的外语人才培养；对外汉语教育；“以商带语”，走出去的企业及其员工负有汉语传播功能
贸易畅通	商贸企业及相关服务行业	商贸语言翻译；城市语言环境	语言翻译；城市窗口行业语言服务	翻译企业、个人；语言技术研发企业（在线翻译、机器翻译等）	发展“互联网+语言服务”，基于云翻译技术的“语联网”；优化国际贸易支点城市及旅游城市窗口行业人员语言服务水平（如交通物流、餐饮住宿、金融通讯等）
资金融通	货币、信贷、投融资、债券等相关管理部门、企业	业务洽谈、政策扶持、标准共建中的“语言转换”需求	语言翻译	翻译企业、个人；语言技术研发企业（在线翻译、机器翻译等）	专门的外语人才培养；对外汉语教育；推动中文成为中间语言
民心相通	民间团体与个人	跨文化交流中的语言消费需求	语言培训以及与此相关的语言出版；语言艺术以及与此相关的语言翻译等；旅游行业语言服务	国家语言文字主管部门；语言培训、语言出版机构；语言文化传播机构；文化旅游主管部门及景区、旅行社等企业	“民心相通”是“一带一路”建设的社会根基，通过推动民间的人文交流，提升沿线国家对中国的了解与认同

注 该表系在作者已发表论文中相关表的基础上整理而成。

对“一带一路”建设中的语言消费研究，首先，可以从对语言消费主体的研究切入，围绕着“一带一路”建设中已经出现和潜在的语言需求，在分类梳理的基础上，选择有代表性、典型性的语言消费主体，运用相应的消费理论与消费者分析方法，对消费者的消费心理、消费动机、消费行为进行调查研究。由于"一带一路"涉及国家多、语言消费主体多、语言消费需求复杂多样，选取具有代表性、典型性、辐射力的调查对象就显得尤为重要，如可以将参与对外经贸合作的中国企业作为研究的突破口，"按图索骥""牵引"出其他的国内外语言消费主体，由点到面，通过逻辑线索清晰地描绘出国内外市场中语言消费的整体图景。

在语言消费研究的框架中，如果说第一层的语言消费研究主要是回答"是什么"，那么，第二层的语言消费研究主要是解决"为什么"，主要研究的是语言消费的"动力机制"问题；第三层的语言消费研究主要是探讨"怎么样"，即语言消费是怎样推动生产的，产生了怎样的效益，同时，第三层的研究还包含一项重要的内容，即国家语言规划与语言政策研究，回应前两个层面研究所发现的问题，探讨语言消费与国家战略的关系，并思考如何从宏观规划与政策层面解决语言消费中存在的问题。

三个层面的研究既层层推进，又首尾相连，第三层直接为第一层所提出的供需问题做出宏观决策的回应，从而在柱形框架的基础上，又构建了一个循环往复的关联系统，如下图所示。

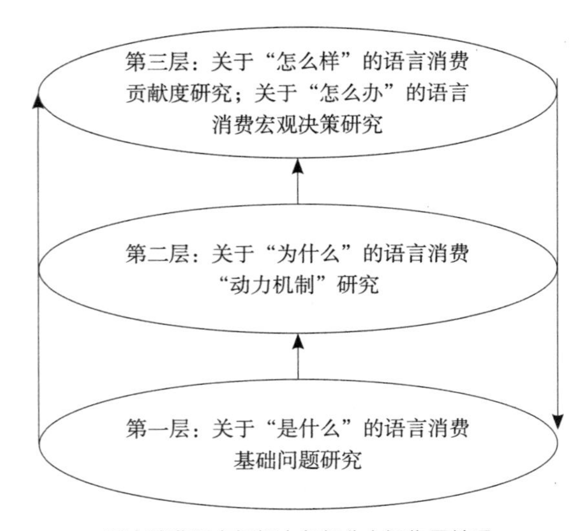

语言消费研究框架中各部分之间作用关系

四、结语

目前，国家统计部门尚未有专门针对语言产业的统计办法和统计数据，但在语言消费需求旺盛、语言产业发展迅速的大趋势下，对语言消费的规模、语言产业的经济贡献度进行测算是非常必要的。

由于语言产业所属各个业态的构成、盈利渠道、经营方式等都不尽相同，所以，我们对各业态采用了不同的测算方法，估算出 2016 年语言翻译、语言培训、语言出版、语言技术、语言测试、语言康复六个行业的整体数据约为 4190 亿元人民币，2016 年我国国内生产总值为 744,127 亿元，语言产业在国民生产总值中的占比保守估算为 0.56%（该统计未包含语言艺术、语言创意、语言会展三个语言产业业态）。语言产业不仅对国民经济具有可观的贡献率，同时也与文化、社会的发展有着双向互动的关系。

随着国内经济、文化以及社会的发展，不仅会推动语言产业的发展，拉动语言消费需求，而且还将促使语言消费需求与行为的多元化、高端化。特别是随着“一带一路”建设的不断深入，语言消费研究也相应亟待跟进。本文尝试提出的语言消费研究基本框架，也还需要在语言产品（服务）生产、消费的发展中得到检验，不断调整完善，以期实现对消费者行为的有效把握，为语言产品（服务）的供给者满足和引导语言消费提供科学决策的依据。

参考文献

［1］后蕾. 对当前“语言消费”现象的几点思考［J］. 南京社会科学，2003（08）：83-87.

［2］黄佩红. 大学生“语言消费”现象分析——以广州为例［J］. 外语艺术教育研究，2007（4）：27-31.

［3］李艳. 语言产业视野下的语言消费研究［J］. 语言文字应用，2012（3）：25-32.

［4］李宇明. 语言服务与语言消费［J］. 教育导刊，2014（07）：93-94.

［5］李艳，齐晓帆. 城市人文形象构建下的行业语言服务能力研究［J］. 文化产业研究，2016（01）：192-204.

[6] 徐大明. 语言服务与语言消费可扩大内需[N]. 中国社会科学报，2012-04-23(B06).
[7] 黄少安，韦倩，杨友才. 引入制度因素的内生经济增长模型[J]. 学术月刊，2016，48(9)：49-58.
[8] 李宇明. 认识语言的经济学属性[J]. 语言文字应用，2012，3(3).
[9] 贺宏志. 发展语言产业，创造语言红利——语言产业研究与实践综述[J]. 语言文字应用，2012(3)：9-15.
[10] 李现乐. 语言消费的个体差异——基于南京服务行业的语言调查[J]. 语言政策与规划研究，2014，2：004.
[11] 赵斌，孙丽丽. 消费行为理论述评[J]. 经济学动态，2009(07)：86-89.
[12] 张卫国. 语言的经济学分析：一个基本框架[M]. 北京：中国社会科学出版社，2016.
[13] 李培林，张翼. 消费分层：启动经济的一个重要视点[J]. 中国社会科学，2000(01)：52-61，205.
[14] 刘凤良，李彬. 消费理论的行为化趋向[J]. 国家行政学院学报，2004(06)：85-88.
[15] 雷五明. 九十年代城市文化消费的特点及其影响因素的调查[J]. 消费经济，1993(03)：24-25.
[16] 李艳，高传智. "一带一路"建设中的语言消费问题及其对策研究[J]. 语言文字应用，2016(03)：94-103.

后记

望着眼前的书稿，不觉中回想起学校第十三届“科学·艺术·时尚”节活动中的一幕幕。从三月份的开会启动到每个细节的实施与落实，倾注了多少领导、老师、同事和同学们的心血；也饱含了外籍专家不辞万里来与我们合作的情谊与友谊；更难忘我们学校领导，在协调雄安日程安排中给予的鼎力支持；校办与宣传部领导预留座位，安排秀场的尽心尽力；民族服饰博物馆馆长面对远方来客，围绕苗族服装的结构，现场展示，以特有的方式讲述了苗族服饰特征为极简，即尽量减少裁剪，体现先人“惜物”情怀与“慎术节用”的造物思想。正如明朝的朱柏庐在其《夫子治家格言》中所言，“一粥一饭，当思来之不易；半丝半缕，恒念物力维艰。”而我自己作为会议主办方论坛板块之一的主要负责人，则结合自己对于传统服饰文化的理解与认识，以《论语》中服饰礼仪为切入点，彰显大国悠久服饰历史，传递中华服饰文化。所有参会的境内外专家学者，实际都在围绕“一带一路”沿途各国的服饰文化，从不同角度剖析解读，使我们对于“一带一路”沿途各国的异域风情，有了全新的感受和理解，仿佛置身于色彩斑斓的服饰文化大观园，畅游、尽享服饰文化盛宴。

这一切的一切，汇聚在这本小书中，虽然因为时间紧，部分学者的论文此次未能及时收录，但

我们依然祈愿这本小书能成为北京服装学院“科学·艺术·时尚”节文化的积淀，为后来者一品昔日的盛宴提供些许回味；期待着明年，我们的新老朋友，学界的大咖与新秀再聚首，共同打造更为丰富绮丽的服饰文化盛宴。同时，感谢学校的各级领导，特别是主管该活动的倪赛力书记的高瞻远瞩与家国情怀，请允许我再次感谢为之付出辛劳的诸位领导、老师和同仁们，包括我们语言文化学院的研究生张佳琪，本科生马铭雪和薛江文燕等。正因为大家的齐心协力，才使我们有理由相信，明年更精彩！

张慧琴

2018年6月1日

樱花东街甲2号

作者简介

李傲君

- 加拿大西蒙弗雷泽大学教育学硕士，北京服装学院语言文化学院教师
- 主要研究方向为教育道德哲学、比较教育、课程设置和服饰英语等
- 主持的课题包括北京市人才引进项目、北服英才项目和北京市优秀人才等
- 热爱学生和教学，2012~2016 年连续五年获得北京服装学院校内“我爱我师”学生心目中最喜欢的十佳教师称号；2014 年被评为“感动北服人物”；2014 年获北京服装学院第十一届青年教师教学基本功大赛一等奖、最佳教案奖和最佳演示奖；2015 年获北京市第九届青年教师教学基本功大赛文史类 B 组一等奖、最受学生欢迎奖、最佳教案奖和最佳演示奖；2015 年获第六届“外教社杯”全国高校外语教学大赛北京赛区商务英语组特等奖等

张慧琴

- 上海外国语大学英语语言与文学博士，北京服装学院语言文化学院院长，教授，中国翻译协会专家会员。美国加州太平洋大学访问学者，英国雷丁大学访问学者
- 研究方向为中外服饰文化差异、跨文化交际和应用语言学
- 主持国家社会科学基金项目和北京市哲学社会科学重点项目等，北京市“长城学者”，多次荣获教学科研成果奖
- 在《中国翻译》《中国外语》《外国语文》《上海翻译》以及《外语与外语教学》等期刊上发表论文 40 余篇，出版有关服饰文化、跨文化交际，服饰文化与大学英语教学融合的专著、译作 10 余部

刘颖

- 北京外交学院英文系翻译与美国研究硕士，北京服装学院语言文化学院商务英语专业教师
- 研究方向为应用语言学、翻译（口译笔译）、美国历史文化
- 参与国家社会科学基金项目，北京市哲学社会科学重点项目，北京市科学委员会项目等
- 曾作为中国常驻联合国代表团的工作人员（大使夫人助理）在美国从事跨文化交流工作。多次独立承担汉译英翻译工作，特别是纪录片《台北故宫》字幕的翻译。翻译出版了《最后一个莫西干人》（*The Last of the Mohicans*）、《非典型美女造型手册》（*Style Evolution*）和《近代汉族民间服饰全集》（*A Collection of Contemporary Han Folk Costumes*）等。先后承担了中国民族学学会与北京服装学院联合举办的"文化遗产与民族服饰"学术研讨会开幕式口译与论文集序言的翻译、大英博物馆代表团参观北京服装学院民族服饰博物馆并与博物馆专家座谈的陪同口译与英文解说、联合国教科文组织主办的世界创意城市北京峰会的各国与会市长参观北京服装学院民族服饰博物馆以及北京服装学院与敦煌研究院举办的"垂衣裳——敦煌艺术大展"的陪同口译与英文解说，以及时装秀的中英文主持等

杨武道

- 华东师范大学博士，北京服装学院语言文化学院副教授，商务英语课程负责人，院长助理。英国剑桥大学访问学者
- 研究方向为文化与应用语言学
- 多次主持北京服装学院商务英语教学改革，商务英语等级考试研究项目等
- 曾在《北方论丛》《外语学刊》《上海戏剧》《内蒙古师范大学学报》与《山西师范大学学报》等核心期刊发表学术论文多篇，出版专著1部，编著1部，翻译文学作品2部

张东晓

- 毕业于首都师范大学英国语言文学专业，获文学学士学位。后在美国阿肯色中心大学，接受管理学本科教育和专业会计学专业教育，获商业管理学士学位。现为北京服装学院语言文化学院商务英语专业教师，曾荣获北京市教育委员会颁发的优秀教师奖
- 研究方向为商业管理、经济及会计
- 在美国学习和工作14年，熟悉美国社会及文化。多次参与各类商务英语跨文化交际活动，协助学校完成诸多重要场合与对外交流的语言服务工作